COAL IS OUR LIFE

To the Yorkshire mineworker

COAL IS OUR LIFE

*An analysis of a Yorkshire
mining community*

Norman Dennis
Fernando Henriques
Clifford Slaughter

TAVISTOCK PUBLICATIONS
LONDON · NEW YORK
SYDNEY · TORONTO
WELLINGTON

First published in 1956
by Eyre & Spottiswoode (Publishers) Limited
Second edition published in 1969
by Tavistock Publications Limited
11 New Fetter Lane, London E.C.4
2.1
SBN 422 73250 8

First published as a Social Science Paperback 1969
1.1
SBN 422 72580 3

© 1956, 1969 by Norman Dennis, Fernando Henriques, and Clifford Slaughter

Printed in Great Britain by Bookprint Limited, Crawley, Sussex

Distributed in the U.S.A. by Barnes & Noble, Inc.

Contents

Acknowledgements

The research which is embodied in the following pages was only able to be undertaken through the generosity of the Nuffield Foundation. We should like to express our great appreciation of the courtesy and assistance we have received from the Foundation.

It is appropriate here to acknowledge the extremely valuable advice and co-operation given us by Professor A. N. Shimmin, C.B.E., under whose aegis the research was initiated.

Without the invaluable guidance, criticism, and comment which was rendered to us by both Professor M. Fortes and Professor M. Gluckman the research would not have achieved the form here presented. We may emphasize, however, that the responsibility for the book rests upon us alone.

Our greatest debt without question is to the people of Ashton. They have neither resented nor rejected our prying into their affairs. Whenever there was an opportunity for co-operation this has been extended to us. Although they remain anonymous, this is their book.

N. D.
L. F. H.
C. S.

NOTE

Ashton is a fictitious name. All place-names which could have been used to establish its real identity have been changed. Large towns in the West Riding have been given their real names.

Introduction to the Second Edition

Opinions differ on the degree of progress made by sociology, since this book was first published (1956), in understanding British society, and in particular the British working class. Consequently there will be little general agreement on the merits of re-publishing this report of the social life of a mining community. The technique of research upon which it was based, that of community-study, has not been either generally used or further developed in the years since 1956. However, this is not the place to review or criticize the tendencies of development of sociology itself, but rather to suggest in what way the findings of a study made over fifteen years ago retain any validity today.

The "community-study" technique, strongly influenced by the monographs of social anthropologists, has obvious pitfalls when applied in societies based on modern industry. By its focus upon the "community" framework as such, this technique will tend to abstract from the societal framework of every level of social life. To express this difficulty in simple form: whereas, in *Coal Is Our Life*, phenomena like the relations between husband and wife, or the nature of leisure activities, are viewed primarily from the standpoint of grasping their interrelation with the forms of activity and social relationships imposed by the coal-mining work upon which the community is based, this emphasis will tend to obscure the fact that each of these particular sets of relationships is extended beyond the community, in both space and time. By itself, the community-study technique provides no way of measuring the significance of its findings against what may be crudely described as these "external" factors.

For example, *Coal Is Our Life* can be interpreted as suggesting that the form of family life found in "Ashton" is somehow a product of the type of social relations enjoined by the particular forms of wage-labour prevalent in the mining industry. However, it is perfectly obvious that the same social form, monogamous marriage and the nuclear family, characterizes many societies and all classes within the same society. The special character of wage-labour and the particular features of coal-mining emphasize certain potentialities of this family form and inhibit others; and the influences upon behaviour within this form derive from a vastly bigger framework than that of a coalmining community (cf. the attempt to begin this correction in Slaughter, "Modern Marriage and the Roles of the Sexes", *Sociological Review*, 1958).

In the sphere of industrial conflict, the same problem arises. While it is true that our account of the miner's work includes a presentation of the economic relationship of wage labour and not only of the immediate work-situation, this economic relationship is never fully integrated into the interpretative framework of the book as a whole. Once again, the community-study technique could not provide any measure of the significance of these "extra-community factors". At an elementary level of explanation, that of the immediate phenomena of strikes and industrial conflict, this limitation of the community-study technique was indicated in an analysis of the large-scale Yorkshire miners' strike of May 1955, just 15 months after the completion of our original research. It was possible to demonstrate that the incidence, mode of development, and consequences of strikes in the community we studied could be explained only at a series of higher levels of social interaction than that of the community (*Sociological Review,* 1958).

While these criticisms do not in any way invalidate the factual reporting of the miner's work given in *Coal Is Our Life,* or of the way in which his work impels the miner into particular kinds of conflict, or of the all-pervading influence of mining on the community's life, they are nonetheless necessary as a corrective to community studies in modern society. We suggest therefore that the re-publication of this book might serve to help teach the limitations as well as the possibilities inherent in the technique we adopted, and to introduce the discussion of methodological problems which remain unresolved in sociology.

Bearing in mind these points, we think that at the same time the reappearance of *Coal Is Our Life* might encourage some re-appraisal of the general reception given to the book in 1956. When investigators report on their own society, they cannot dismiss the problem of the effect of their own published conclusions, and their responsibility to those about whom they write and to whom their reports are addressed.

It should be remembered that our study began some five years after the nationalization of the coal industry. Since then three times that length of time has elapsed. In the early nineteen-fifties strikes and other manifestations of conflict were generally regarded as a survival of pre-nationalization conditions, and therefore destined to disappear, once the "psychology" of the miners adjusted to the new reality. Our explanation (even with the limitations indicated above)—that the recurrent conflict and the attitudes of miners were the outgrowth of the actual economic relations and working conditions under the given form of nationalization, rather than any survival of outmoded habits—was ill received. Our analysis of the way in which the inner-union relationships were adapted to the continuation of this conflict of interest, and of the way that this adaptation was actually legitimated and at the same time obscured by the prevalence of the assumptions of a real change, was generally rejected. We suggest that the last fifteen years provide an adequate test for these controversies.

Similarly, our presentation of the cultural poverty and isolation imposed on the working-class community, of the material factors which effectively limited the value of "welfare" provisions and changes in social mobility through education, was generally seen as a somewhat morbid preoccupation with the miners' past. When we described the many ways in which women were oppressed by the relationships imposed by the miner's work and dependence on wages, we met the same type of criticism. Here again we suggest that, while the continuation of the economic boom and "prosperity" through the nineteen-fifties and early sixties nurtured such objections mightily, 1969 provides a more sobering vantage-point. Even in the matter of wage-levels and consumption, we believe that the facile assumptions of middle-class observers of the life of the working class are questionable at every point, but, at the more general level of the whole cultural horizons open to the working class, we do not consider that the questions we raised have even begun to be considered. Perhaps one statistic will serve to illustrate the effect of the kind of factors which we described in 1956: in 1969 the chance of a working-class girl entering University is one in 600.

Our book was dedicated to the Yorkshire mineworker, and this was not the result of a fit of sentimentality: the Yorkshire miner is perfectly well able to speak for himself, and he will yet do so. But he faces a situation now, in 1969, which demands new answers, different from those which he has given to his problems in the past. Here the relation between an investigation like ours and its material becomes extremely complex. The investigator's *responsibility* to those about whom he writes is not an abstract principle, but arises in very concrete forms. Here, for example, we may ask: does *Coal Is Our Life* point towards the situation in which the Yorkshire miner now finds himself, and indicate a way for him to confront it? In other words, did the book isolate and emphasize those factors which indicated the real life-course of the community, or not? We suggest that our report confirms the opinion of most miners themselves, already in 1952-4, that the promise of security, prosperity, and a new life flowing from nationalization and the welfare state was a lie, was justified, and must be developed and built upon rather than discarded or exchanged for the ideological dream-world of "affluent societies", "embourgeoisement", and "the institutionalization of conflict", not to mention that mirage mistaken for a discovery by two intrepid explorers in the desert of industrial sociology, "the withering away of the strike".

The mining industry, or certainly a large part of it, has been condemned to death. Yorkshire is not one of those coalfields which has already been virtually completely closed, but more and more of its pits face extinction. We do not mourn the death of a dangerous and health-destroying industry as such, any more than do the miners themselves. Insofar as this decline is part of the abolition of arduous manual labour then it is a step along the path to human freedom. But in the social relations actually existing it is not yet any such thing. Social relations do not adapt to the cultural basis of progress automatically or according to the course of reason. As a worker, the miner has only one basic right: to sell his labour-power. As a member of the working class, he exercises that right under conditions established through a century and a half of organization and struggle. For miners today, the death of their industry means that the heart is torn from their communities. There is no overall planning of new industries or of training or of education for leisure; there is no more than marginal provision for economic security. We did not paint the mining community in any *couleur de rose*, but this community without the mine and mineworkers is in danger of becoming merely an aggregate of socially isolated and culturally condemned human beings. Miners themselves have always considered that the social system considers them as fit only for the scrap-heap when they cannot be employed. The miners' union leaders have accepted the closures imposed by the Government and the National Coal Board, never questioning the criterion of "economic" applied to each pit; they have insisted only on the level of redundancy payments. Their horizon was never higher than the narrow trade union consciousness so typical of the "official" British labour movement, nourished as it has always been by the smug provincialism of the Fabian Society.

However, this situation is not unalterable but, like all social situations, problematical. Relations inside the miners' and other trade unions, and in the economy and political system as a whole, are undergoing important changes, under the whiplash of necessity. The future of the miners does not rest with the raising of hands on the National Coal Board. The very fact that miners do *not* live only in communities, but as part of their class in all its economic and political relations, means that the alternatives available transcend what can be observed or deduced from the life of the community. Paradoxically, the miner's fate as a mineworker and as a member of a mining community has made it necessary for him, if he is to survive, to look beyond the prospect to which the planners would like to confine him. It is in this sense that we say the miner has still not said his final word.

Is it possible to omit these problems in discussing the circumstances of the reprinting of this book? There will be many who think so, but we find it difficult to see how the sociology of our own society can fail to place at the centre of its considerations the matter of its own actual relation to the forces which it analyses. Whatever its intentions, in one way or another, it becomes part of those forces so soon as it is communicated.

Centre for Multi-Racial Studies   Fernando Henriques
University of Sussex   March 1969

CHAPTER I

Place and People

YORKSHIRE might be described as a microcosm of England in that every type of region in the country from remote hill farms to industrial towns is represented. The West Riding itself is one of the major industrial bastions of Britain. In this area from the inception of the industrial revolution until the present time industry has been the keynote of life for the majority of its inhabitants. Some towns have become famous for a particular product such as wool textiles or machine tools. Others have supplied the driving power of industry in the form of coal.

The town of Ashton, which is the object of the present study, since its collieries were opened in the second half of the nineteenth century has played this latter role. Its name will not be found in any atlas or gazetteer since it is our invention, but the place is a very real one. Ashton, with a population of nearly 14,000, lies in the centre of the Yorkshire coalfield and, as far as our observations go, is fairly typical of the area.

Ashton was once a rural village with an ancient history stretching back to the "Domesday Book". Its agriculture remained relatively undisturbed until the first of its collieries, Manton, was opened in 1868. That year decided the town's destiny. From that time its inhabitants have depended for their existence on coal.

Physically the growth of the town has been dependent upon the development of the collieries as the table on page 12 shows.

The dominant feature of the landscape is the spoil or slag heaps. There is no point in the town from which they are not visible. Houses and mine-workings crouch under their shadow. To the observer the spoil heap is the physical symbol of work and life in Ashton.

Housing illustrates the ideas which employers, individuals, and authorities have had in this respect during the last century. The

TABLE I

Population of Ashton Urban District correlated with the
development of the collieries in the district

Date	Population			Colliery development
	Total	Males	Females	
1801–61	600–700			
1871	2,265	—	—	Manton Colliery opened 1868
1881	5,901	3,171	2,730	First shaft at Ashton Colliery sunk 1877
1891	7,528	4,152	3,376	Ashton Colliery shaft deepened to second seam 1885
1901	12,093	6,578	5,515	Two new shafts sunk at Ashton Colliery 1892–4
1911	14,374	7,698	6,676	New shaft sunk at Ashton Colliery 1910–13
1921	14,842	7,789	7,053	
1931	14,955	7,934	7,021	
1951	13,925	6,975	6,950	

mineowner's mansions, the suburban semi-detached, the local
authority housing estate, the squalid rows of back-to-backs, with
communal privies, the terraced rows of artisan's houses, all have
their quota in the town. You can walk in five minutes from 1870
to 1955. The pollution of the air is such as to reduce clothes, houses,
and streets to a drab uniformity.

Back-to-back houses today are in a minority. They number
about 100, and were built by the original colliery owners for
miners. Accommodation consists of a living-room-kitchen, one
bedroom, a box-room, and a cellar. In this part of the town
streets are still unpaved. There are blocks of shared lavatories in
the middle or end of the row. Their existence, however, does much
to reinforce the outsider in his opinion he formed on passing
through–"Ashton ? Oh, that dirty hole." This is an opinion very
commonly expressed by outsiders. To some extent it must influence
the opinion of the people of Ashton in regard to their town.

Other houses dating from the same period, *circa* 1870, are of the
artisan type. As they were built on the then edge of the town

large yards and open spaces are common. Most of them have two bedrooms, a kitchen, and a sitting-room. Although greatly superior to the back-to-backs their disadvantages are obvious. There are no bathrooms, and the W.C. is shared with the house next door. The large yards do not merit the name of gardens for as often as not they are used primarily as dumping grounds for refuse of all kinds.

But it is the years 1891–1911 which give Ashton its characteristic stamp. Ashton today contains approximately 4,000 houses. During the period 1891–1901 one quarter of these were built together with nearly all the public buildings. In 1891 there were 1,323 inhabited houses, in 1901 there were 2,237. Many of these houses can be classified as of the 'superior' artisan type. With three bedrooms, a kitchen-living-room, their own yard and separate W.C. they represent the best of pre-1914 building. Apart from appearance it is the actual siting of the houses which is characteristic. The general tendency was to build up both sides of existing roads with the result that at this period Ashton resembled an urban thread laid across the countryside.

Between 1911 and 1953 nearly 1,300 houses were built. By far the great majority of these are local authority houses or estates built for the National Coal Board. These estates have been built away from the main roads, sometimes admirably sited, and to a great extent hidden from the casual observer. Thus it can be said building of this period has not altered the superficial characteristic stamp given to Ashton in the years before the 1914 war. These houses are clearly the best in Ashton from practically every point of view. The average size is three bedrooms, a combined kitchen and living-room, and a bathroom and W.C. In addition garden space is good.

Table II shows the development of housing in Ashton correlated with the growth of the population.

It is clear that standards of housing in Ashton vary considerably. There is the same variation from cleanliness to utter filth in the interior of the houses. Again, homes will reflect the relative prosperity or poverty of the owner. But it cannot be said that there is any necessary connexion between bad housing, dirty homes, and poverty. There are homes on the new housing estates which are

TABLE II

Year	Population	Inhabited houses
1871	2,265	349
1881	5,901	1,087
1891	7,528	1,323
1901	12,093	2,237
1911	14,374	2,570
1921	14,842	2,868
1931	14,755	3,296
1951	13,925	3,856

excessively filthy but in which there seems to be plenty of money for the household, and there are back-to-back homes which are impeccably clean. Behaviour in the home is apparently not governed by the type of housing but by other factors.

Ashton is predominantly a working-class town owing its development to the growth of its collieries. The latter have drawn people and houses around them, the main pit is almost in the centre of the town. But the collieries have exercised a centripetal influence in other ways. Most of the men in Ashton are miners. The cohesive results of this fact are well known. First, there is the inapplicability of the miner's skill to the other trades. Secondly, there is the long history of acrimonious disputes for which the coal industry is notorious. Common memories of past struggle have undoubtedly helped to bind a community such as Ashton.

While the nature of the work and the history of the industry in Ashton have thrown the men together in this way, they have exerted an opposite or centrifugal influence on the women. The coal industry provides no paid work for them. In an area where there is no alternative they have to do without it. Ashton, however, lying as it does in the coal zone of the West Yorkshire industrial region, is within reach of alternatives. Fogarty points to the absence of paid employment for women in the coal area.[1] He bases his analysis on the returns of the Labour Exchanges which

[1] M. P. FOGARTY, *Prospects of the Industrial Areas of Great Britian.* London, 1945.

show that while in the whole of Great Britain and Northern Ireland in 1935 there were 25 insured women workers to every 100 insured men, and 55 to every 100 in the twelve largest West Yorkshire towns, yet in Calderford, the Labour Exchange which is adjacent to Ashton, and which town, though similar in industrial structure, is not dependent to the same extent on coal, there were only 6 insured women to every 100 insured men. While this is true, it would be a serious error to miss the point that women were working outside their own high ratio area, in areas where the ratio was lower, and to ignore the effects on social life resulting from the fact that from the time that the town ceased to grow until the Second World War, there was a considerable surplus of men between the ages of 15 and 29.

The 1951 census shows that there is now a slight surplus of females between the ages of 15 and 29.

Not only can the *character* of the industry be shown to disperse as well as bind, but the *history* of the industry in Ashton reveals the same dichotomy. In 1935 one of the collieries, Ashton Main, was permanently closed. Ashton Main miners applied for work at other collieries outside the town. On the one hand Ashton owes its being to colliery development, and the integrative tendencies of the miners' trade helped make the town a community. On the other the centrifugal tendencies resulting from the absence of paid employment for women were also at work. The fact that one of the collieries was closed was an additional consideration of this nature. The collieries, then, are seen to have exerted influence in both directions.

A comparative survey was made of a nearby town, Fullwood, and all the points raised here, apart from the closing of one of the collieries, apply equally well to it.

There is, however, one outstanding difference. Ashton is within easy reach of three substantial towns, and there are frequent bus services to each. Castletown with a population of 23,000 is a 5-minute journey away, Calderford with a population of 43,000 barely 15 minutes. Ashton has never been physically isolated. Fullwood on the other hand is 10 miles from Barnsley, the nearest town to exceed it in size. The ecological significance of this can be

TABLE III

Age distribution—Ashton

	1911		1921		1931		1951	
	m.	f.	m.	f.	m.	f.	m.	f.
Single	4,833	3,777	4,630	3,834	4,473	3,453	3,429	3,373
Married	2,649	2,617	2,874	2,880	3,143	3,150	3,543	3,590
Widows and Widowers	216	292	285	339	319	419	315	300
Age group								
0–4	1,066	1,112	910	902	708	698	722	698
5–9	931	998	928	972	768	809	639	590
10–14	796	812	929	872	800	772	557	538
15–19	753	481	806	603	823	589	486	510
20–24	726	492	627	553	755	541	453	529
25–29	622	485	554	558	678	566	610	570
30–34	570	496	522	458	610	529	524	576
35–39	543	446	492	461	491	495	575	564
40–44	496	358	445	412	477	426	518	527
45–49	331	268	443	378	428	404	506	476
50–54	264	210	356	289	398	356	464	473
55–59	220	172	292	215	341	299	362	327
60–64	180	142	189	127	282	225	286	274
65–69	122	116	151	113	192	135	263	247
70–74	54	61	86	73	98	79	174	201
75–79	17	21	44	47	62	64	88	109
80–84	6	14	15	20	16	27	46	47
85–89	1	2	—	—	8	6	13	7
90–94	—	—	—	—	—	2	1	—
Totals	7,698	6,686	7,789	7,053	7,935	7,022	7,287	7,263

demonstrated by comparing the effect of the difference in location upon two spheres of local social activity.

The two activities examined were club life and shopping habits. While there are sixteen registered clubs in Fullwood there are only eight in Ashton. The combined membership of the former is over 10,000 (10,180) and the combined membership of the latter under 7,000 (6,844).

These figures appear to show that Ashton has a looser hold upon the leisure-time activities of one section of the men. Another set of figures seems to show that the Ashton housewife in going about the business of shopping goes farther afield than does the Fullwood housewife. For while there are only seven shops selling clothing in Ashton (the shops included here are tailors, outfitters, and drapers), in Fullwood there are nineteen. Ashton and Fullwood are not identical in all respects except that of location, and, therefore, it cannot be said with confidence that the observed differences in social behaviour result from location. But both towns are dependent on mining to a similar extent; both developed in a similar fashion, though Ashton's development was earlier. Both are in the same industrial region, viz. on the borders of the West Yorkshire and South Yorkshire coalfield; and both are about the same size, though Fullwood is somewhat larger (with a population of 18,000 against Ashton's 14,000). They differ, in the manner described above, that is, in the proximity to towns of the one, and the remoteness from towns of the other. This last circumstance might therefore be deemed to exercise a strong centrifugal influence on social life in Ashton.

These two tendencies, the centripetal and the centrifugal, help to determine the 'community' for the citizen of Ashton. Clearly what is the 'community' for one person may be much less than the community for another. Dispersive and cohesive influences impinge with varying degrees of power on different sections of the community. While for some women, working in Castletown and spending their time in Calderford, the 'community' may be very attenuated, for an old miner, Ashton may be very nearly the all-inclusive home of his being.

GETTING A LIVING

A description of who gets Ashton's living involves first of all a comparison between the number of those who are available for employment, and those who are dependent upon them.[1] It is convenient to regard the former as those in the age group 15–64.

[1] In the description which follows we have compared the statistics of the years 1911 and 1931. At the time of going to press the 1951 Census Occupation and Industry Tables had not yet been published.

The latter is composed of children who because of State regulations are forbidden to work, and old people who because of custom or illness, either are not allowed to work or are under no pressure to do so. In this respect Ashton is living in a very real sense on the past. It has today an abnormally large percentage of the population of working age because in the past it bore a heavy burden of dependent young people. Between 1911 and 1931 the population of working age as a proportion of the total population increased from 57·4% to 64·9%. Other things being equal, therefore, Ashton would have been considerably better off in 1931 than in 1911, for out of every 100 people 7 who had been dependent in 1911 were in 1931 of age when they could earn their own living. The explanation of this, of course, is not only that the product of the higher birthrate of the earlier years moved into the ranks of those of working age, but also the birthrate was lower in the later years. This is clearly a temporary advantage, and Ashton has, with the rest of the country, stored up unavoidable trouble for the future, since the present abnormally large proportions of people of working age will move, in the future, into old age. Thus a large number of people will have to be kept in a leisure to which they feel entitled, by a working population which has been recruited from generations less numerous than those they replace. It is unavoidable because the working population which will have to keep them is already born. A higher birthrate now would only worsen the situation, by matching the heavy burden of the old with the heavy burden of the young.

The population of working age, however, is not the same as the working population. This immediately excludes one large group, the housewives, who are of working age–who work, but not for a living in the sense of being gainfully employed. The increase in the proportion of the population of working age has resulted in a similar change in the proportion of the working population. Whereas in 1911 38 out of every 100 were working for a living, in 1931 there were 42. Of these 42, 7 or 16% were females, an increase of 6% on 1911. Paid employment for females appears to have become more common in Ashton. But the figures show that relatively few Ashton women go out to work. In 1936 in Great

Britain and Northern Ireland as a whole, there were 25 insured women workers for every 100 insured males. In the twelve leading towns of West Yorkshire itself, there are 55 insured women for every 100 insured males.

The question of the age of those working has already been touched upon. In 1911 for every 100 males aged 15–64 there were 106 working. There are now 104. The two outside the 15–64 age group who were working in 1911 and who are not working now are children who would in 1911 have secured, as a reward for ability, permission to leave school at 13 or even earlier, and old people who were obliged to work because of the pressure of a necessity which is no longer felt to such a degree.

The conclusions to be drawn are two. First that the healthy adult male in Ashton is, as he was in 1911, almost without exception engaged in the activity of getting a living. Among the women, however, no such consistency of tradition is apparent. Compared with the other places mentioned, the proportion of women working is low, but compared with Ashton in 1911 there has been a significant increase. Secondly it can be seen that while the working population from the point of view of sex is getting wider, from the point of view of age it is getting narrower.

The quantity of the effective working population is determined among other things, by the phenomenon of unemployment. This has been such an important fact in the life of Ashton that a full description of its causes and incidence in relation to Ashton is demanded. It is a good example of the influence exerted on Ashton by virtue of its membership of an international economy, of the influence of a particular industry, of a particular district, and finally of influences internal to Ashton itself.

It would be inappropriate to discuss at any length the causes of the depression of 1929–33. It is sufficient to say that the social fact of cyclical unemployment[1] was experienced in Ashton in those years and it affected the whole town. Ashton has been alternately prosperous and depressed throughout its existence due to international fluctuations in trade. It was born, so to speak, in the travail

---

[1] Cyclical unemployment, i.e. by a general decline in the demand for goods on an international scale.

of the slump of 1876–9 and experienced unemployment and low wages, as the normal course of the trade cycle brought unemployment and the depressions of 1883–6, 1893–4, 1902, 1912, 1921, and 1930–3. It experienced also the intervening booms, and in the present post-war period all look apprehensively to the state of the American economy in the knowledge that to some extent they are dependent on it.

The town was affected again by the fact that its livelihood depended upon the state of the demand for coal. This not only fluctuated periodically in the above manner, but also suffered a steady decline after the First World War, and a steady rise after the Second. The reasons for the decline are well known and will be merely mentioned. First there was the decline in demand caused by the more efficient use of coal; secondly there was the competition of other fuels, such as oil and hydro-electricity; and thirdly there was the development of overseas mines.

It is not possible to measure the exact effect of this in the form of structural unemployment[1] except for the years 1937–9 for at all other times Ashton has been included for these statistical figures with a dissimilar area. Some idea of the magnitude, however, can be gauged by using the figure for the unemployment of mine-workers in the whole of the West Yorkshire area including Ashton.[2] Calderford, Castletown, and Bousfield Exchange cover 28,000 of a total of 33,000. In some of the years covered, of course, cyclical unemployment is superimposed upon the structural.

Unemployment in the coal industry has been quite different in kind and degree from the two so far discussed (i.e. cyclical and structural), both of which are caused by a deficiency of effective demand. It is caused by temporary maladjustment due mainly to local facts such as the working-out of seams. A good example of this is the closing of Northfield Colliery (Lanarkshire) in February 1954, which threw 350 miners out of work. This sort of unemployment has not been experienced in Ashton since the war.

In addition to the experience of Ashton which sprang from its membership of an international economy, and of a declining

---

[1] Structural unemployment, i.e. caused by the decline in the demand for the goods of a particular industry which is not part of a general decline.
[2] See table on next page.

TABLE IV

Unemployment among mineworkers in
West Yorkshire 1928–36

Year	% of mineworkers unemployed
1928	21
1929	27
1930	28
1931	26
1932	32
1933	57
1934	39
1935	38
1936	29

TABLE V

Unemployment in Ashton Labour
Exchange area – 1937–9

Year	% of all insured workers in Ashton unemployed
1937	48
1938	23
1939	13

industry, there was, thirdly, the less important effect of its membership of the political community. The structural unemployment of the coal industry was in a sense written into the law of the land by the Coal Mines Act of 1930. Under Part I of that Act, the output of every owner was subject to regulation by quota. This meant that the expansion of the labour force of an efficient group of collieries was forbidden. It also meant that instead of the full pressure of unemployment being brought to bear to encourage the movement of its labour into industries which could employ it, a colliery worked for part of the week, until its quotas were fulfilled, and the labour was unemployed for the remainder of the time.

The factors which generate unemployment have a differential impact according to local conditions. Thus unemployment was not as high in West Yorkshire as in South Wales and not as high in Ashton as in some of the Durham towns. The reason for this can be found in Ashton's position *vis-à-vis* the markets for its products. South Wales and Durham looked mainly to the severely contracted export markets. Whereas Ashton with the rest of West Yorkshire supplied mainly the industrial needs of the West Riding, and the domestic needs of its large concentration of population.

Finally there are factors peculiar to Ashton itself, which, given the circumstances, affected the manner and degree of its unemployment experience. The most important of these was the fact that the Barren House Seam at Manton Colliery ceased to be an economic proposition at the current prices. The result was that in August 1935, 560 men and boys were dismissed and the seam closed. Shortly afterwards, the remaining employees were dismissed, making a total of about 650, and the colliery permanently closed.

The state of trade at the time being given, there were two reasons why Manton was closed. One was geological. The colliery was first opened in 1868. Its exploitation followed the course of all but the modern collieries which use the retreating long-wall system. The coal most easily won was won first, and the colliery's history was, therefore, one of diminishing returns. The second reason was a social one. The workmen at Manton Colliery were 'well-organized'. In the spring of 1935 there was a dispute with the management over the percentage of 'dirt' (i.e. stone) being sent to the surface among the coal. The Yorkshire Mineworkers Association sanctioned a ballot at Manton, and there was a big majority in favour of a strike.

Here two interesting points are brought to the fore. The first demonstrates Ashton's integration into the life of the district immediately surrounding it, the second its independence of it. The manager of Manton had said that "too gloomy a view was being taken of Ashton's position. The Manton Colliery was permanently closed but in due course it would be possible to absorb practically all the men displaced. . . . With the means of transport we have today, there should not be much difficulty in Ashton men going to work in other collieries." This somewhat facile reassurance was vindicated to some degree, and by May 1936, 75% of the Manton men had been absorbed elsewhere in the coalfield. Ashton's independence of the surrounding district is illustrated by the fact that the Manton miners who went to work at collieries in the neighbourhood often experienced the greatest difficulties in adapting themselves to the custom of their work-places.[1]

The account of unemployment has already made it clear that Ashton is a coalmining town. In 1911 there were 5,001 occupied males, of these 3,790 (that is 76 out of every 100) were employed in the coalmining industry. By 1931 when the number of occupied males had risen to 5,481, the number occupied in mining had fallen to 3,728, or 68 out of every 100.

These miners do not all work in the town. There are 2,300 employees at the only two Ashton collieries, Ashton and Vale.[2] Of these, 600 of those at Ashton came from outside of the town. For the 3,700 Ashton miners, therefore, there is work in the town, for, at the most, 2,300. This means that out of every three miners in Ashton one at least must travel to a colliery outside of the town to his work.

No other industry employs more than a small proportion of Ashton's people, and in this too, though there have been changes, the position today is substantially that of 1911. Ashton is still a coalmining town. Its other industries, with the exception of clothing, are ancillary. Consider first the distributive trades. In 1911 there were 232 Ashton persons getting their living in this way, of whom 174 were males and 58 females. This represented

---

[1] "At N. where I worked, after six months on the dole and on the union," a contract worker said, "there was a 'butty-man' in charge of each team of ten men or so, five on each shift. The butty-man used to get the pay for the whole team and pay it out himself. All the men from Ashton insisted on drawing their own wage. There was nothing but trouble. One day I snatched the paynote from the butty-man and I was so annoyed when I saw he had been paying us short that I knocked him into the waste and shouted at him. Luckily nothing happened—my mates stopped me, and next day, the butty-man came and told me not to worry, he wouldn't say anything."

This is comparable to the experience of Scottish miners. See H. E. HEUGHAN, *Pit Closures at Shotts and the Migration of Miners*, pp. 55–7, Edinburgh, 1953.

[2] In November 1953 there were 1,755 employed at Ashton Colliery. The average number of employees at Vale Colliery in 1952 was 653.

4·2% of the occupied population. In 1931 6·1% of the occupied population were in the distributive trades. These figures do show a relative rise in the lighter industries which was characteristic of the period in Great Britain as a whole. They also show that, by comparison with other towns in the West Riding, for Ashton people the change was a small one. The proportion of insured workers in the distributive trades in Leeds in 1935 was 14%, in Keighley it was 8%, and in Morley it was 7%.[1]

Another of the ancillary industries is transport. In 1911 there were 70 males engaged in this industry. All except a handful worked for the railway company. By 1931 the number working on the railway had dropped to 54, but 194 were now employed in road transport. Of these, 187 were men and only 7 were women. There is no means of telling how many of these worked in Ashton. In 1954, however, it is known that only 137 of the 240 employees at the local omnibus depot were Ashton residents. The 1954 figures show too that the proportion of females has greatly increased, there now being 182 males and 58 females.

In both the years 1911 and 1931, the occupation which most Ashton girls and working women entered was that of domestic service. In 1911, 169, that is, 34% of all occupied females were in domestic service. The only other occupations which claimed more than a total of 20 were food dealers and general shopkeepers (58), dressmakers (54), teachers (46), and midwives, sick nurses, etc. (21). In 1931 the proportion of domestic servants was still 30% of all occupied females, the remainder being engaged mainly in commercial and financial occupations (15%), makers of textile goods (11%), and professions (9%). In view of the fact that Ashton itself is predominantly working class, it is probable that nearly all those classified as domestic servants must have been earning their living away from Ashton. What this restricted choice of occupation and particularly the concentration on domestic service meant can perhaps be better envisaged by using a case history to illustrate it.

[1] These figures are not quite comparable as the Ashton figure is that for all workers, but since many of those in distributive trades would be uninsured owners, the disparity serves only to emphasize the point being made. On the other hand, the categories included in the figure of those occupied in distributive trades for Ashton are broader than those for the other towns.

"After being educated in the usual manner at what are now the Primary and Secondary Modern schools at Ashton, this informant left school at 14, in 1939. She immediately went into service in Huddersfield, after her mother had been told of the position by a friend. There she received 10/- a week and her keep. She hardly ever went out, and getting very few letters from her mother, after a few weeks was so homesick that she left her job and returned to Ashton. She had saved 30/- out of her wages of £2 and bought each of the family a present. Then hearing from a friend of hers of a job in Manchester, she went again into service with the friend. Her employers were Russian Jews, and she disliked the food cooked in oil, so after 6 weeks she decided to leave. She had saved no money so she went to the police station, where she had a ticket home purchased for her. She then went to Windsor, where her eldest sister was living and after two jobs there, as a shop assistant and the other as a cinema usherette, she came back to Ashton. Then followed a year's employment at a Leeds factory to which she travelled daily, work with the N.A.A.F.I. which again meant leaving home, and finally work as a conductress with the local bus company. It was here that she met her prospective husband, who was a driver in the same firm, and she left work at 19 when she was expecting her first baby."

In 1944 Ashton was scheduled as an area which required facilities for female employment. Therefore, there are now two small clothing factories and a very small macaroni factory. The latter employs less than 10 women and can be ignored. The largest clothing factory employs a total of 180 of whom 13 are men. Of the 167 females, 30 come from areas outside Ashton. The other factory has 30 employees, of whom 29 are women, and of these, 26 are Ashton women.

This analysis of the industrial distribution of the population does not complete the description of the business of getting a living in Ashton. But it shows that for the average inhabitant of the town coal is the means through which he gains his living.

CHAPTER II

The Miner at Work

INTRODUCTORY

THE fact that over 60% of the male working population of Ashton finds employment at the local collieries has significance at three levels. Ashton's families have a common fate determined by virtue of their similar relationship, through a wage-earning husband, to the coal industry. In that miners are wage-workers their social relations have much in common with millions of others in Great Britain and over an even wider area. At the same time, this similarity is modified by the particular conditions and history of the coal industry as such; at this level Ashton has social relations in common with other mining communities. Finally there are factors derived from the particular nature of Ashton itself, including particularly the local availability of alternative employment, the local market for coal, and the actual physical conditions of coal-getting in the area. Therefore it is proposed in this chapter to consider the relation between work and life in terms of three questions.

(a) What is the range of effects on the life of the miner and his family of the fact of being of the 'working class', in the sense of the above definition? This will involve a specific description of the status of the wage labourer in our society; it is not, of course, possible to ignore the question of the effects of nationalization on miners in this respect.

(b) How is the life of the miner and his family specifically affected by the fact of being of the *mining* section of the working class? Here the particular character of mining as an occupation, in its physical aspects, its differences from other types of work, and the place of miners in the history of the working class, will be prominent.

(c) In what ways do 'local' factors in the sphere of work and employment give an individual, unique bias to the life of Ashton mineworkers?

Before these three sets of problems are embarked upon it is worth remembering that we do not suppose these factors to work independently or autonomously; they work in and through each other. To take but a single example: the basic geological facts of coalmining as an industry have made for the agglomeration of smaller communities than have other aspects of Britain's industrial growth. Now, therefore, miners live in communities of a different character from those industrial towns characterized by diversity of occupation, social class, and varied social and cultural amenities. This is the basis for a certain autonomous community life in such small towns or villages as Ashton. One further qualifying note is necessary; a discussion of work in its relation to life bears upon almost every field of social relations mentioned or baldly stated for the moment. These are left for specific and detailed elaboration for other chapters, such as those on the family, leisure institutions, or trade union organization. Moreover, we are only temporarily leaving aside the question of how these other aspects of life have themselves an effect on work and working attitudes.

THE MINER AS A WAGE-WORKER

It has already been pointed out that the miner enters into the process of production by selling his labour power to an owner of capital. Just as the employer's ownership of plant, of means of production, would be of no value without the availability of labour, so, given this ownership, the worker needs the facilities of employment offered by the employer.

There are certain important facts about such a relationship between a class of men and their work.

The position of the worker in modern industry

The role of the worker is not to direct production, it is to put himself at the disposal of the employer for a certain period of time. As a consequence of his labour power being bought by the employer, the product of his labour is alienated from the immediate producer, the labourer, and is appropriated by the employer. The workers receive as the price of their labour power, wages which they use to purchase other commodities, the means of life which enable them to reappear on the scene with their labour power, and to reproduce themselves. It is in this sense that the family is the economic unit of society and of the working class, with the head of the family as its representative in production. With certain insignificant exceptions, in modern industry this economic relationship exists stripped of any direct human relation between worker and employer. The relation is one between labour and capital. It is rare for a worker ever to meet his employer. The worker does meet some of the executive staff and supervisors of the work; this part of the staff has the function of directing and controlling the worker's labour. This fact of the worker being only the repository of a commodity (labour power), bought by the employer, and the fact of obvious contradiction between the social, co-operative production, and the private appropriation of the product, are the fundamentals of the worker's position in our society. From them are derived numerous consequences including the superficial phenomena of 'the attitude of men to their work' about which we hear so much.

It is repeatedly suggested that trouble in industry and frustration among workers can be attributed to the monotony inherent in the operations allotted to most workers in modern industrial processes. In the first place, it should be said that the monotony entailed in the particular processes carried out by the workers is often over-estimated, and there is certainly no indication that in those industries where monotony is greatest, 'industrial morale' is correspondingly low. The facts of accidents, absenteeism, and strikes in mining – one of the least monotonous of jobs – provide a sufficiently cogent example. It is doubtful if the monotony of the job has any great independent effect on morale, and it would seem to be quickly over-ridden by stronger influences. Yet there is a sense in which every worker suffers 'monotony'. It is not the monotony of the operations he carries out, considered in their concrete aspect, so much as the tendency for his work to be directed and controlled entirely from outside himself. Different occupations, from that of the collier to the assembly line worker, allow for varying degrees of freedom within limits. But they have this in common; the plan of production is independent of the worker, and it incorporates his labour just as it does raw materials, machinery depreciation, and every other constituent of the final product.

In addition to this absence of scope for initiative and creative effort, there is another consequence of the worker's status which is responsible for similar effects in behaviour and attitude towards work. Again these effects appear as 'monotony' and therefore give rise to the facile judgments already mentioned. The wage-worker does not go to work by choice. He arrives at a time set by the requirement of the enterprises. The time at which he leaves is also dependent on the needs of the job. Should he wish to spend a certain amount of time in some non-working activity during the hours when he would normally be at work, he cannot as a rule arrange things so that his work is completed quickly through extra effort, so as to leave him the time he requires. Certain situations may arise where this becomes possible; for example, in a period of shortage of labour and high wages the worker can exercise free choice in the use of some of his time without endangering his security. Generally speaking, however, and considering the life-experience of the wage-worker, his maintenance of life depends on regular fulfilment of the labour-wage contract with his employer, and acceptance of the conditions involved.

It is clear, therefore, that although there is in one sense a reciprocal relationship between the owner of capital and the owner of labour (the worker) in terms of his life-process, the worker experiences his tie to the enterprise as a continual and binding necessity. When a man receives his wages every seven days, and these are on the whole not a great deal more than enough for comfortable survival, he is *bound* to his work. By Sunday night the collier who starts work at 6 a.m. on Monday is not enjoying himself with the same abandon as he did the night before. By Wednesday, three hard days may have made him tired and dispirited and he consoles himself only with the remark that at least the back of the week has been broken. Favourable conditions of wages and the state of the labour market might make it possible for a few men to stay away for one or two days in a week. But the general fact, the fact which abides week by week, month by month, and year by year, is that money, the means of life, is secured only by a regular commitment to the workings of the enterprise to which the worker is attached.

The attitudes arising from this position are not surprising. One or two examples may be quoted. In Ashton the working men's clubs hold a weekly draw with the first prize of £30 or more; most members take odds of 600 shillings to one with the local bookmaker that their number will be drawn, so that the total winnings can amount to about £60. For a prizewinner in this draw to turn up with his workmates on Monday morning would occasion a great deal of surprised comment. Again if a man's absenteeism is explained by someone who knows he has won the draw, or that he is expecting a dividend from the football pools, no further questions are asked. It is thought natural to take the opportunity of not going to work when such a windfall occurs. This is only the most pointed example of a general attitude. Mr. J. A. Hall, for many years president of the Yorkshire Miners' Association, in a public statement in reply to critics of absenteeism in Yorkshire mines, argued that any man working in mining would take the opportunity of 'buying his leisure'. The very small group of habitual absentees, as anyone in a mining area knows, includes men who would hardly think of going to work if they happened to have thirty shillings in their pockets on Monday morning.

A second example concerns the voluntary Saturday morning shift in the mines. The appeal for such a shift to be worked was made on the basis of the dire needs of the national economy for increased production. The following discussion is typical (abridged).

"J.F. 'Coming in on Saturday?'
"A.S. 'No. Five days is enough for anybody.'
"J.F. 'Oh, so you're not bothered about getting some extra coal out for the country?'
"A.S. 'I suppose that's why you come in on Saturdays.'
"J.S. 'Is it...! We come in for some extra brass and that's that.'"

These examples (there are a great many others) show that the miner views his work as a necessity, without which sufficient money would be available to ensure only the barest necessities. His work is the opposite of freedom, as he sees it, and yet no freedom is possible without it.

Some jobs may be more interesting than others, conditions of work may vary a great deal, certain economic tendencies may put the worker in a temporarily advantageous position *vis-à-vis* the employer. All these factors can modify, but only modify, conclusions about the general, compulsory, everyday tie of the worker to his work. A group of miners waiting to ride out of the pit had an interesting interchange of views. One man preferred the afternoon shift (2–9.30 p.m.) to others because it gave him the mornings free to dig his garden, and he got plenty of sleep. Another liked the day shift (6 a.m.–1.30 p.m.) because the day's work was early over and done with and he could go out in the evenings. A third liked the night shift best because it made a long week-end, not starting work till Monday night. But the whole group acclaimed the man who clinched the discussion with, "If you're anything like me, you don't like any bloody shift."

If it is not the worker who initiates, controls and directs his participation in industry, neither is it, in the worker's own experience the actual owner of the industry who does this. In every industry there is a staff of supervisory grades and managers. There are also technicians and other planners but rarely are these encountered by the workers themselves. It is the managers, foremen, deputies, etc., who appear on the scene of the worker's experience to direct and supervise their work.

Ideally, perhaps, such officials have the sole function of promoting efficiency in production – the smooth running of the enterprise concerned. But it is impossible to escape from the fact that in reality they are obliged to deal with problems of labour discipline. Just as the worker's other social relations seem opposed to his work rather than smoothly integrated with it, so the managerial officials appear to be enemies in many ways. They must take sides whether they like it or not. Although the Coal Mines Act of 1911 makes it quite clear that the colliery deputy is essentially a SAFETY official, he is traditionally regarded as the 'Bosses' man' and day-to-day life in the pit sees a constant struggle for advantage between the deputy and the workmen. In other industries the equivocal position of gangers and charge-hands, who are placed both in the group of workers and outside of it as a part of the managerial staff, shows the impact of these social considerations in what could presumably be a purely 'efficiency' arrangement.

Industrial relations and 'class-consciousness'

That workers in all industries do see management or employers as their opponents cannot be doubted. The very fact of the enormous trade union organizations that have been built up in this country over two centuries is sufficient witness to that fact. Whether or not the workers are class-conscious in the political sense is another question, but that they are certainly class-conscious in the limited sense of their view of the immediate relations between worker and employer. Certainly some employers run their factories in more benevolent and far-sighted ways than others, and many Labour and trade union leaders have seen as the aim of the working class the ideal employer co-operating wisely with the needs of the workers![1] For the most part, however, workers have learned the lesson of their own history, that business is business. They see it as natural that the employer wishes to make profit out of their work; they accept in the main the idea that this is a sign of a man 'getting on in the world'. But their aim also is to make money, and for this reason their relationship with the employer is one of struggle for the division of the spoils. The argument and the doubts on the issue are summed up in the words of an ex-miner, B. A.

For twenty years he worked as a miner, but by winning some thousands of pounds in horse-racing he was able to establish a bookmaker's business, employing a small staff. When a collier, he had become well known for strong left-wing opinions, and he still condemns 'the capitalist system'. He is worried about his relationship with the men who now work for him. "I remembered

[1] Such an outlook seems to have been most common in the old craft unions; cf. evidence before the Royal Commission on Trade Unions, 1867; and T. J. DUNNING (1799–1873), *Trade Unions and Strikes; Their Philosophy and Intention*, London, 1860.

how the bosses used to treat the men," he said, "and I thought I'd try and rule by kindness. But I soon found I was being swindled; they thought I was soft and took advantage of me. You see they'd got so used to ordinary employers that they thought this was just a good chance to get the better of me. So I had to change my methods and now I have to rule them with an iron hand. I don't like doing it but I find I can't make it pay any other way."

The National Coal Board has met with difficulties similar to those of B. A. Mining in particular, of course, has been an industry characterized by long and bitter antagonism between workers and employers. National Coal Board officials complain bitterly that any change they introduce is greeted with mistrust, even if its advantages to the men would seem to be immediately obvious.[1] When the National Coal Board or any other employer talk about the need for 'economies' the workers expect attacks on their wages. If cubic capacity is introduced for measurement of coal production in place of the old weighing system, suspicion immediately comes to the fore and the men question how much they are going to lose. Scores of examples could be given of the prevailing idea among workers that any suggestion emanating from management, since it is designed for greater profit, is likely to be some underhand attack. It will benefit one side or the other; coincidence of interests is unthinkable and they cannot conceive of the boss being philanthropic.

The working man does not necessarily see the picture as we have described it. His class-consciousness is not simply an extension of his antagonistic relationship with his own employer to all those in a similar category. He thinks not in the abstract terms of social and economic relations which we have used, but in a more concrete way. For example, his pride in being a worker and his solidarity with other workers is a pride in the fact that they are real men who work hard for their living, and without whom nothing in society could function. It has always been a favourite point of Socialist speakers to tell their audiences how indispensable they, the workers are in comparison with the 'parasites' of the upper classes. The

[1] This type of phenomenon is not, of course, confined to the field of industry. There are examples of colonial areas where a tradition of bitterness based on past experiences tinges the attitudes of the people to every action of governing power.

'parasites' are marked off essentially by the fact that they get a living without working. Nothing provokes more anger among working-men than to hear from non-manual workers exhortations to increased effort. In 1953 Sir William Lawther, himself an ex-miner, made a speech of this nature which was fully reported in the BBC news bulletins. Mineworkers in the public houses of Ashton greeted him either with groans of boredom or with pained anger, "Why the hell doesn't he try it himself?" The reactions of two women, one a miner's wife, were identical—"Oh, why can't they leave the bloody miners alone. It's all right for him sitting in his office. He wants to go down and have a try for himself before he opens his mouth."

In conditions where the worker feels himself at war with his employer, what is his attitude towards hard work? His main concern is not whether production is greater or less except in so far as this affects his wage-packet. Again there may be exceptions, where other motives enter, such as war-time emergency. There is another argument used by working men against greater effort in work. It is generally understood that economic crises are crises of overproduction; workers will be unlikely to boost production if they know that their firm has a dwindling number of orders. Older miners will express concern if the stocks of coal in the pit-yard begin to pile up.

This matter of the attitude of the miners to production-drives and efficiency brings us back to the characteristic of insecurity which is so important in the life of the wage-worker. Whether or not a man works is dependent on the capacity of owners of capital to cater for him. Thus unemployment often appears to the workers as a deliberate and personal act of will; many miners regard the chronic unemployment of the last depression as the planned outcome of Conservative policy. And so the workers' real insecurity, consequent on fluctuations in national and international markets, is reinforced by his view that individuals are capable of visiting misfortune upon him. Apart from this permanent possibility of unemployment and deteriorating conditions,[1] the worker must

---

[1] This may seem paradoxical, in that the miners have experienced fifteen years of full employment, and some of them have comparatively high wages, but we are considering essentially the life-experience of the worker, not just a segment of that experience.

also find that insecurity results from the very fact of working for a weekly wage. In certain industries, because of material conditions or because of piece-work, it is difficult to know whether one's wage will be good or bad. Illness or an accident can reduce one's earnings and one's assurance of employment to the minimum. From all this springs the strong support for social security among the working class. They strongly depreciate the opposing arguments; to argue that social security detracts from individual initiative and self-confidence and may lead to laziness and abuse shows, they would say, a lack of understanding of a worker's life situation. Both points of view are the understandable rationalizations of class-conditions and interests.

We have said that the working-man tends to have a concrete view of his relations with other classes, i.e. he sees and remarks in the main the outward signs of the fundamental relations. He thinks of the fact that the other classes do not perform manual work rather than of specific social and economic categories and relations. He confirms this view by a consideration of the spending and consumption patterns of the different social strata. In general his class is the class that has to work for an amount of money that will give enough to live on, with a margin for enjoyment. The other classes are marked by possession of a comparatively large supply of money, either 'in the bank' or assured at definite intervals.

This concrete conception of the marks of class is the key to the workers' attitude to 'getting on' in the world, or what is called upward social mobility. In the sense that we have discussed the class relationship, i.e. structurally, according to position in the economic framework, there is an ever dwindling chance of mobility for the worker. The tendency is for greater concentration in industry, for the cutting-out of competition, and thus for the end of the 'little-man-who-made-good' days of private enterprise. In the basic industries such as coalmining and iron and steel, the possibility of a man becoming 'his own boss' is nil; it does persist, with important effects on class-consciousness in industries such as building. A good example of the individual impact of these tendencies is L. W. (aged 30):

"His father was a miner and determined that his son should work elsewhere. L. was therefore apprenticed as an engineer with a Leeds firm, to which he travelled twelve miles daily. On his return from military service in 1945, in the sellers' market then existing, he used his gratuity to set up in business as a contracting welder. In this he was quite successful for just under two years, at which point he discovered that certain technical innovations were spreading in the trade which made it impossible to compete with larger firms who could do jobs more quickly and cheaply. Nor could he buy the new equipment, for this entailed the purchase of plant which called for a larger capital than he could command. He had by now married and had two children. After working for himself he did not intend to work at his own trade for an employer; in any case the post-war golden age of easy employment at good piece-rates in engineering was fast disappearing. By 1953 L. W. had entered the mines as a trainee after trying various jobs. He could not quite make up his mind whether he should train for a good steady job in mining or try again 'on top', i.e. at some job on the surface."

It appears as if the workers cannot attain the status and way of life of the upper classes by assuming a similar position in the economic system. But in fact does anyone want to achieve it? The answer is a result of the concrete rather than the abstract nature of the class distinction. Indeed the fact that ordinary workers talk far, far more about class distinction than they do about class struggle or other aspects of social class is significant in itself. Propaganda about inequality and injustice has a naturally strong appeal to working people;[1] in all sorts of ways these are the marks of their station in life. Since they can no longer conceive of 'getting on' in the old Samuel Smiles sense they seize on the most conspicuous outward characteristic of the class difference, and this is spending-power, the possession of wealth.

In his work, a man's concern, over and above the minimum requirement of 'holding the job down', is money. Over the past fifteen years money earned at work has raised the miner a certain amount in the social scale within his own class—his family is no longer marked by the poverty of the inter-war years. But 'winning the pools' is the great vision of ending worries and giving a

[1] Features of this type of news such as the colour bar in South Africa and in this country receive considerable attention in Ashton and other mining villages. A nearby village organized in 1952 a public meeting with Seretse Khama which enjoyed a very large attendance.

man the chance to decide his own destiny, to be no longer at the mercy of all that his job represents. Money gained in such ways holds out the possibility of breaking down barriers to improvement for the worker and his family.

Obviously enough, the channels of real transformation, such as the treble chance coupon, will carry only the exceptional and fortunate few, and the rest will in the main produce only short-lived or modest improvements. It is a fact that wage-workers remain wage-workers until they are 65. The ideals of behaviour, the good things of life, in short the cultural ends of the society in which they live, remain for most men a vision only, in the glossy magazines, the newspapers, on the cinema and the television screen, and in the lives of a few people whom they will never encounter. These 'cultural goals', to borrow expressions from Merton,[1] are not equally available to all participants in the culture because of the inequality of the 'institutionalized means' placed at their disposal.

In terms of national economy and society, the inhabitants of Ashton are part of a class-divided society. It is interesting to note here that Ashton itself, far from being a microcosm of that national framework, is representative of only one part of it.[2] To all intents and purposes the inhabitants of Ashton are all of the working class. In relation to the stratification of our society they are all in the same category. In this, Ashton is typical of mining villages. Whether or not this characterization of the wage-worker's position is as true for the miners in 1953 as it was for miners in 1939 or for workers in non-nationalized industries at the present time is a problem that will be clarified by a closer examination of the particular features of the work of miners.

[1] R. K. MERTON, *Social Theory and Social Structure*. Glencoe Free Press, 1949.

[2] This is the principal reason why it will be impossible to arrive at systematic conclusions of the type reached in social anthropology, i.e. that devoted to more primitive and homogeneous societies. ". . . the anthropologist can choose for himself a locality 'of any convenient size' and examine in detail what goes on in this locality; from this examination he will hope to reach conclusions about the principles of organization operating in this particular locality. He then generalizes from these conclusions and writes a book about the organization of the society considered as a whole." (E. R. LEACH, *Political Systems of Highland Burma*, 1954.)

THE MINER'S WORK

Although older miners insist on the increasingly easier nature of the colliers' work today, it is invariably said by miners that pit-work can never be other than an unpleasant, dirty, dangerous, and difficult job. A description of the different types of work in the mines will be useful in a discussion of attitudes to work, and the relation between work and life; if broad differences emerge between the work today and past conditions, then we will expect these to be reflected in the lives and attitudes of the different generations of workers.

Mineworking in Ashton itself is fairly typical of British mining, the degree of mechanization not being exceptional in any way. Approximately 53% of all underground mineworkers in Britain are 'contract-workers', i.e. they are engaged on piece-work.[1] This percentage includes all those working at the coal-face, and men engaged on development work, which is usually either the making of roads, i.e. tunnels in the rock, or the opening out of new coal-faces. Those at the coal-face consist of 'colliers' (the term varies for different parts of the country – hewers, fillers, etc.), 'machine-men', 'drawers-off', 'rippers', 'pan-turners' (or panners), and a few others. The clearest view of the work of each type of face-worker is gained if we follow the daily production cycle, which is divided into three shifts.

THE PRODUCTION CYCLE: WORK AT THE FACE

Our coal-face may be of any length from 50 to over 300 yards, and in Ashton collieries, may vary in height from 3 to 5 feet.

Machine-men. A team of four or five 'machine-men' operate a coal-cutter on one shift. These cutters are usually worked by electricity, though some in the area are still run on compressed air; their function is to make a cut of two or three inches width either at the bottom, the top, or the middle of the seam, cutting into the face by means of a long jib to a depth of from four to six feet. The whole work of the rest of the production-cycle is to 'fill off', i.e. to send out of the pit, the coal cut, and to make the face safe

---

[1] In Ashton the figure is slightly lower. (48% in 1953.)

THE MINER AT WORK 39

and ready for recommencing the cycle in one day's time. The cutting machine displaces a certain amount of coal in the form of fine dust, and this has to be filled on to the conveyor-belt, which runs a few feet behind the face, and parallel to it. Between themselves the members of the cutting team divide the jobs of driving the machine, operating the belt and the panels which control power, cleaning away dust (which may amount to 15 or 20 tons per shift), and making the roof safe. Machine-men must be strong, quick-thinking, and very competent workers. There are obvious hazards and difficulties in operating and manœuvring a machine weighing many hundredweights and using considerable electric power – particularly in a limited space such as the coal-face offers. In addition it must be remembered that the operation performed by the cutting jib creates a tremendous vibration and tension in the coal-face and in the roof and floor. All this necessitates a combination of skill and strength with the maximum co-operation within the group.

'*Drawers-off*'. Working more or less at the same time as the machine-men are the 'drawers-off'. Again this is a dangerous and heavy job, consisting of removing timber and steel supports from the waste. The waste is the area left empty, apart from supports, by the previous working of the coal-face. Thus on a 100-yard face with a cut of 1½ yards an area of 150 square yards is left empty each day. This is left to be supported by specially constructed stone 'packs' at definite close intervals, and the drawers-off must salvage the valuable daily supports, where this is possible; where steel props and bars are used, this is patently more important. Pressure both from the roof and floor often leaves the props very firmly embedded, and even where pressure has only been slight the tension can be very deceptive.

An inexperienced man trying to salvage timber can cause very great danger to himself and others. With experience a pitman can judge to a nicety from which end to approach an awkward bar or girder, or in what direction the released bar – say a piece of steel a yard or two long – will spring. Besides 7-lb. and 14-lb. hammers and picks, the drawers-off use a sylvester-chain – a simple lever device which enables them to keep at a distance from their work.

This type of work is part of the daily routine of every face-worker, but only the drawers-off are concerned solely and continuously with this job.

The colliers. On the following shift, usually the 'day shift', the largest team of workers in a pit, the colliers, arrives. The colliers' job in the $7\frac{1}{2}$ hours shift (perhaps 6 hours at the face itself) is to move from face to conveyor-belt some 10 to 15 tons of coal: this he does with a shovel. To get the coal to the floor, whence it can be placed on the belt, he uses a pick, sometimes a mechanical one; he is aided, in addition to the coal-cutter, by explosives. The face is divided into 'stints', i.e. stretches in a face 4 feet 6 inches in height of some 8 to 10 yards. Each collier is allotted one of these which he must clear to the depth cut by the machine,[1] placing props and bars every yard along his stint. Thus his safety depends on himself. He must guard not only against falls of roof, but also against the face of coal falling towards him. This last is responsible for many means' accidents some of them fatal.[2]

From start to finish of the shift the collier keeps up a hammer and tongs rhythm, broken only by some fifteen or twenty minutes for 'snap'.[3] Because of the intensity of his work he wears only 'bannockers' or cloth shorts, boots, knee-pads and cap with lamp. His shift is spent entirely on his knees, in which position he comes to use a shovel, pick, and hammer more expertly than he can when standing on his feet.[4]

One or two men on the previous shift, working behind the machine-men, will have bored holes for shotfiring[5] though often

[1] In speaking of the 'depth' of the cut, reference is to the extent to which the coal-face is penetrated by the cutting jib each day – three feet or four and a half feet.

[2] In 1953 there were 118 deaths, 382 reportable accidents, and 35,403 accidents altogether due to falls of the roof at the working face; 43 deaths, 283 reportable accidents, and a total of 18,333 accidents due to falls of the face or the sides at the working place. (*Ministry of Fuel aud Power Statistical Digest.*)

[3] There is not a lunch or tea break at a definite time, as in other industries, but twenty minutes can be taken at an agreed time for a sandwich or two and a drink of water.

[4] In some parts of the country miners regard it as essential to stay on their feet wherever possible, cf. *Once a Miner*, N. HARRISON, Oxford, 1954.

[5] Shotfiring – in every stint holes are bored to approximately the depth of the cut. In this the deputy or shotfirer places an explosive charge; he then 'stems' the hole tight with clay and all men at the face retire with him while he sets off the charge with an electric battery. In this way the coal is broken and can be brought more easily to the ground for removal. In 1951 new regulations were published for safety with shotfiring, but figures as to their effect are not yet available.

colliers bore extra holes for themselves, if the coal is difficult to get. Sometimes the firing of shots takes place during the colliers' shift but it is a growing practice for teams of shotfirers to complete the operation before the arrival of the colliers. Besides making the work smoother and cutting out delay, this method obviously makes for a greater degree of safety. However, there is still a remarkably large number of accidents from shotfiring, remarkable since this is one sphere where rigorous enforcement of official rules could reduce the figures to nothing. The *Mines Inspectors' Report* for 1953 shows that in that year 15 deaths and 36 reportable accidents were directly attributable to shotfiring, not including falls of roof resulting from shotfiring, and other explosions.

It seems in this case that safety precautions have not kept pace with the technical changes. Between 1940 and 1951 there was an increase of over 57% in the amount of explosives used in mining. At the same time, output had declined since 1939 by some million tons a year, so that while in 1940, 7·11 tons of coal were produced per pound of explosive, in 1951 this figure was only 4·28 tons. The cost for the collier of a great deal more explosive power to his elbow in coal-getting has meant an increase, directly from that cause, in serious casualties of 88 above the 1940 figure of 204. This tendency in shotfiring is exceptional to the general tendency of a decline in the rate of serious accidents with technical advance.

Sometimes the colliers' work may vary but little from day to day over some months. On the other hand he literally never knows quite what to expect in the way of working conditions. Natural variations of many kinds can confront him. On some days the roof or floor will move inwards with tremendous force. In more serious cases this can close in a whole stretch of face; very often it causes falls of several tons of stone, and endangers the system of supports which has been constructed. Men come to know those sounds in the roof which suggest a fall, and sometimes a whole team of colliers must scurry into the roadways which are more permanently settled and better supported. It is common knowledge that many colliers opposed the introduction of steel supports because they had grown so used to detecting danger from the creaking of timber props and bars. Besides facing the danger of large-scale movements of the roof the collier must be prepared for danger from the actual condition of the stone immediately above him. Sometimes a roof will be solid, i.e. formed of a good thick stratum of hard stone, but on the other hand it may be formed in such a way that pieces of stone crack from the rest of the roof and fall to the floor. These minor falls at the face may vary from pieces of some hundredweights to long, sharp, slivers; both have their dangers and both tend to take the workman by surprise if he is not constantly watching the roof. Different seams have their characteristic roof conditions, and even then there are considerable variations from day to day. Old men who have been out of the pit for a score of years and more can remember in intricate detail the types of 'top' which can be expected in the various seams at different stages of their development.

There are other variations in natural conditions which affect the collier's work, though none of them so seriously as roof conditions. The coal itself, according to its formation, may have a tendency to fall inwards on to the working space, or to slide in certain directions once it has been dislodged. From a face where it was advantageous to break in from underneath and work upwards and inwards, a collier may move to another where such a technique would involve danger. Again, the variation is not only between seams, but may be greater or less from one day to any other. Only by developing a capacity for continual adaptations of techniques and concentration to these variations can the collier maintain efficiency and safety over any continuous period in his working life. The statistics for accidents show that no collier avoids accidents altogether, and a high percentage have serious accidents at one time or another.[1] Coal dust at the face should also be mentioned, for although it does not immediately affect the work of the experienced collier, it has devastating effects along with stone dust on the long-term health and fitness of miners. Dust suppression has until recently been extremely backward in this country, but

---

[1] In 1953, out of every 100,000 employed 57·1 met with fatal accidents (a total of 401). This is less than half of the rate when the Coal Mines Act was passed in 1911. Reportable accidents (fractures, dislocations, and other serious personal injuries) in 1953 amounted to 1,958 or 310 in every 100,000, similarly reduced since 1911. The total of all accidents was 233,870.

rapid strides are being made by the National Coal Board. The incidence of ill health and death from pneumoconiosis is of terrible proportions.[1]

More exceptional conditions, though they are by no means rare, can hinder the work at the face. Water, either on the floor of the workplace, coming in from the roof, or in the coal itself, is one of the worst of these. Even when it does not occur in large quantities, or is pumped away quickly, it impedes efficiency and safety by softening the roof, making tools and timber difficult to handle, hampering the collier's speed and freedom of movement, and, of course, making the coal heavier. In addition, men working in such conditions are likely to suffer in health. Another familiar feature interfering with the smooth running of production at the face is the occurrence of geological faults in the coal seams. A fault is a break in the stratification of the coal deposit and since this will not be conveniently symmetrical with the working direction of the face, it causes severe complications and delay, and is fraught with the danger of falls. It may take some weeks for regular day-to-day production to recommence on a face which is working through a fault. Naturally all other workmen at the face besides colliers are affected.

Other factors act on the work of the collier besides 'natural' or physical ones, and they depend on the efficiency of the colliery as a whole. The conveyor-belt which runs along the face is ideally in continuous movement throughout the shift, but in practice it often stands still. This is either because of its own mechanical breakdown–breaking of the belt, inefficient maintenance, short-circuit in electric motors, or over-loading–or because conditions are not right for loading the coal at the end of the belt–perhaps the transport system is suffering from breakdowns, or the distribution of coal-tubs to the different faces is not well managed. For these and many other reasons the colliers at the face may find that they cannot go ahead continuously with their filling, with the result that they have to hurry unduly in the last two hours, or stay over-time. Again much will depend on the work of the other men in the production cycle, particularly the borers and machine-men.

[1] In 1952 the number of new pneumoconiosis cases was 3,397.

If the boring is skimped, i.e. if there are very few or badly spaced or insufficiently deep holes bored, then either the colliers must complete the job themselves or the shotfiring will not have the effect it should. If the machine-cut has left a few inches of coal above the cut sticking to the roof, or under it on the floor, then the colliers are left with awkward and sometimes dangerous extra work.

We have mentioned these various complicating conditions of the colliers' job, not simply to emphasize the difficulties he faces—though this serves to distinguish them in some ways—but principally because these conditions have repercussions on industrial relations. Colliers are normally paid on a piece-work basis, reckoned on agreed price-lists. This price-list is based on an assumption of normal working conditions, and for every circumstance which is abnormal the collier fights for concessions in his wages. All contract-workers work on similar principles, and we shall later discuss the problem of wages in greater detail.

Colliers and 'team-work'

The machine-men work as a team; in a different way, so do the colliers. Every collier must definitely complete his own stint each day. If one or more on the face do not, there is danger that the whole face may lose a day's production and wages. Alternatively, a workman on another shift will be sent to complete the stint, and he must be paid 'off the colliers' note', i.e. out of the total earnings, based on the output of the face, of the group of colliers.

If one man is meeting with difficulties the collier next to him or another man who is a quick worker will help him out, but if any individual consistently falls behind and needs help, the colliers exclude him from the team and make it clear that he must go 'on the market', i.e. take any job that is going when he turns up for work instead of having a regular contract. Before this step is taken it would be usual for the men on the face to approach him and say that they are considering such a move. Thus they 'give him a chance' to achieve a satisfactory standard.

A very common phenomenon is for men to stick together through many different contracts for years on end, sometimes for a score of years and even a working lifetime. A whole team often moves from a worked-out face to a new one, and with a few changes may last as a team for a dozen years. There tends to be a core around which the team is built, some of the additions staying on, others drifting to a new team, being rejected, or finding employment at another colliery. The strongest and most permanent alliances are between pairs of men, though sometimes three men will stick together for long periods. These groupings affect the day-to-day work. Friends will work next to each other, help each other out in filling and timbering, in certain conditions even work their stints jointly. There are occasions when the team for a new coal-face is made up of say two groups of three who are unacquainted one with the other, with a few single additions to make up the team; whether or not a harmonious combination is built out of such elements depends on the extent to which the smaller groups are closed and on the personalities of the individuals concerned as much as on their skill. As a rule a good workman is accepted by the rest of the team, but if he does not also fit in socially it is doubtful if he will stay long in the team; he must be a good miner, and his workmates must feel they can trust him. In teams of colliers containing two or three different nuclei of close friends there exists either good-natured rivalry and chaffing ('kidding') between the groups or just tolerance so long as efficiency is maintained. Colliers cannot afford to allow such differences to develop into antagonisms.

The teamwork of the colliers is not so much joint activity but the performance by each individual of an allotted task each day. Should he fail to consistently carry out this task, he imperils the future regularity and security of good work for his mates; he stays in the team only if he is efficient. Within this team which operates the whole face, small cliques develop within which mutual aid is practised. The team of colliers, with this system of mutual dependence, is the hub of the social structure of coalmining.

The panners

The collier leaves his workplace cleared and safe for the faceworkers on the next shift. A team of men, all trained pitmen, has the job of moving forward for the next day's work all the equipment at the face. The steel pans and rollers which support the conveyor are brought into line and the belt broken and re-planted on the reconstructed pans. In addition the 'panners', as they are called, must make safe the space previously taken up by the belt and its supports. To do this they build wooden baulks ('chocks') and stone packs at regular intervals; these packs are constructed solidly of stone, and entail the lifting by hand and shovel of two or three tons of stone, which is, of course, much heavier than coal. Other heavy jobs for the panners are the turning round of the machine to begin the next day's cutting and the moving forward of the engine or gear-head which drives the face-belt. The work of the panners is most efficiently done by dividing into pairs. Each pair will take, say, the moving of two steel pans and building of one pack. The other jobs will be divided between the different pairs. Like the colliers' team, that of the panners varies in number according to the length of the coal-face. It is a team in a similar sense to the colliers', but responsibility is divided more between pairs than between individuals. As amongst the colliers, lasting groups and partnerships develop.

The rippers

Simultaneously with the panners the rippers are at work. Leading to each face two 'gates' are kept open as roadways, usually one used for conveying coal (the loader-gate) and the other for timber (tail-gate, etc.). These gates are made up to 20 feet in height and solidly supported by girders and covering material. Along the loader-gate there will usually run a conveyor-loader-belt, on which the face-belt empties. The loader-gate leads on to the main transport system where the coal empties into tubs to be carried to the pit bottom (i.e. the bottom of the winding shaft).[1] It is the daily job of the rippers to keep their gate – there is one for each team – up to the workings of the face. In the main this means ripping stone from the roof to the required height and width of the gate, leaving no loose or dangerous stones and setting

[1] The latter is a description of an inadequately mechanized or 'average' system of coal-conveying, just as is our description of the production cycle. In any one pit there will be differences of detail of one sort or another, but our aim is to give the main outlines of work.

girders and covering material under the roof and sides they expose. It is now usual for arched girders to be used. The tools used are no different from those of the collier – pick, hammer, and shovel, with the aid of shots fired in the stone. Although a team of six rippers in a loader-gate must divide the jobs of ripping, boring, 'packing' the gate-sides, moving forward the gate-belt, timbering and girdering, the nature of their working area means that they are practically all the time within sight of each other. If one man is ripping with an ordinary or pneumatic pick he must be quite sure that he is not endangering men working on the floor within a very small range.

The moving of the gate-belt cannot be undertaken until the roof of the gate is made quite safe. An 18-foot-high steel arch-girder cannot be lifted, held up and fixed safely without a skilful co-ordination of effort from preferably more than two men. Ripping is a dangerous job because of the wide expanse of roof exposed, and this solidifies the ties between the members of the team; there is no room in a ripping team for a slap-dash or slow-witted workman.

Ripping contracts usually supply the highest wage-packets in the pit, comparable to those of the machine-men.

Besides face-workers, there are men on contract work occupied with the ripping of new roads, the expansion of old ones for transport purposes, and the opening of new faces. The conditions in which they work are more settled and less complicated than those of face-rippers, and more than two men could not comfortably work together in the conditions of this 'back-ripping'. Over a period two men come to understand each other and find out the best division of labour between them, so that it is usual for such partnerships to last several years, the men moving from one contract to another. Often the men on such work have been face-workers and are well enough known among the other men in the pit, but it is worth noting that they are more isolated than any other category of contract-workers. Often they will work a whole shift without seeing anyone other than the shotfirer and the haulage hands who bring them tubs once or twice a day. They are paid according to the progress made each day, and so their success depends entirely upon themselves, excepting only the proviso that tubs are available, but even this is not always so severe a handicap as it is for the collier. In their work these ripping contractors are thus more independent and secure than face-workers.

Men developing new coal-faces form more shortlived combinations. They must make inroads in the chosen seam from existing roads, and prepare the length of face for the introduction of mechanical equipment for the working of the coal.

Once again this group must include skilled workers, who know the requirements of a new coal-face, but it may also include men who are 'working their passage' to the new coal-face job. They will accept the often heartbreaking and difficult job of opening a new face at a very ordinary wage on condition that they are given a regular stint on the new face. In these intitial stages of working a face the collier has to make his own working space. He often has to shovel back his coal three or four times before he is in a position to completely clear it. Ventilation is poor until the face is open from one end to the other, and the absence or distance of good roads means he may be short of timber and of tubs to clear his coal. Men engaged on this opening-out of new faces are sometimes not on piece-work but are paid a daily rate above the minimum day-wage.

TYPES OF CONTRACT-WORK, AND WORKER-EMPLOYER RELATIONS

All the jobs we have described are done by men or groups of men with regular contracts to complete their particular job every day. This attachment of a man to his own work is of some importance. All miners recognize the desirability of such a contract, and in any pit there are several experienced contractors who by their skill and reliability are in a position year by year to negotiate good contracts with the management. The varying types of team-work and working conditions we have described are important in differentiating groups in the pit. Men working in large teams, such as the colliers, whose conditions vary over short periods and whose wages are, therefore, not by any means a certainty, tend to show great solidarity vis-à-vis the management.

In fact it is noticeable that most mining disputes are derived from dissatisfaction among the coal-face workers. However, the more independent pairs of men who have 'back-ripping' contracts, in conditions which alter very little from day to day, tend to set more store by maintaining good relations with the management and undermanager. In a remarkable number of cases men on such jobs as these have the reputation of being 'bosses' men' and 'tale-tellers'. Another interesting phenomenon is the number of union committee men, in Ashton, and nearby collieries at any rate, who hold down such jobs. The following example is typical: two older men (aged 56–60) both prominent union committee members, and one of them a candidate for nomination by the Labour Party for Parliamentary elections, have been working on the same contract for three years. They are expanding an old roadway by a process of dinting or lowering the floor. They made their contract with the manager when the job began at 60s. for every yard advanced in the road. By clearing 1½ yards a day they earned £2 5s. per shift each, or just over £13 weekly. At the beginning of the contract their work was worth this amount; they would fill perhaps 20 tons of stone daily, set timber, and lay rails for the roadway. But after some time the work became easier, and they often filled only 6 or 7 tons per day, and this was more easily worked. Their contract still applied, of course, and their wages did not suffer, although they did little more than a man on the day rate of 25s. a day. It was months before the undermanager, whom they knew very well from mutual membership of local clubs and committees, came round to the job and quietly suggested that in the interests of fair play perhaps some revision of the contract would be considered, and this was agreed. No such delay would have been tolerated by the men themselves had the work grown consistently worse instead of better. Again, had the contract been one for colliers or any group without such exceptional relations with the management, it is certain that the management would have sought immediate rectification.

Face-workers, in fact, find themselves in a day-to-day tussle over wages. The management seems to cultivate good relations with a smaller number of contractors, those working in smaller

4

groups, and apparently regards relations with face-workers, especially colliers, as naturally antagonistic. Of course, there are many exceptions. Skilled machine-teams often include men with good relations with the management, and no doubt managers are keenly aware of the vital function of this small group in the maintenance of production. There are also a few examples of colliers who compromise with management rather than see their interests as mutually antagonistic. In practice, the existence of such a group of men means that they are cultivated so that the management always has on hand a number of 'reliable' workers who can be called upon to carry out emergency jobs, replacing absentees, staying hours and even full shifts overtime, and even being on call from their homes to the pit. The ordinary collier sees himself only in the capacity of doing his job well, getting as good a wage as he can, and then being free. When he sees a number of well-paid men who put themselves at the disposal of the management, who are never without a good job, and who are often even very friendly with the colliery officials, he will begin to suspect that these are 'bosses' men'. Despite what we have said about union branch committee members often coming into this category, it should be emphasized that in the main, posts on union branch committees are filled in the first place by the representatives of the coal-face workers. Among the other contractors one tends to find older union men, who in their younger days had the reputation of firebrands, but have found a satisfactory *modus vivendi*, and either drop out of union activity altogether or exert a 'moderating influence' on the younger union representatives. We shall see this process of differentiation and its effect on trade union organization and militancy in a later chapter, and before we can appreciate it fully a description of the remaining groups of workmen in the pit is required.

THE 'MARKET' AND THE 'DAY-WAGE' MEN

The market-men

Between the regular contract-workers and the 'day-wage' men, mostly haulage-workers, stand the 'market-men'. In an industry with a much higher than normal accident rate, and in

which face-workers are the most common absentees, both voluntary and involuntary, there needs to be a reservoir of skilled men.
On every shift the undermanager, or the overman, will have the
job of employing a number of such men – market-men. This
category is composed of men who have had contracts which have
worked out, of young men not yet established in a team or a
partnership, of men who have not long since arrived from another
colliery, and finally of men who are not regarded as sufficiently
reliable in their work or perhaps are consistent absentees.[1] The
market-men sit in the pit-bottom or at a pit meeting station on
the road to the coal-working until the overman has a clear picture
of the attendance on each job. In place of every absentee, where
possible, he sends a market-man. Some market-men are capable of
almost any job and many have five different jobs in a week;
others may be regular substitutes in teams where absenteeism is
high. An example is G. A., a young, recently married man who
had just completed his training for face-work. For six months he
worked with a team of panners whenever they had a man missing.
Although absenteeism in this team was spread over five or six men,
it became more and more necessary for one man, D. H., to need
replacing, for he was working only two or three shifts a week : at
the time he was under the stress of domestic difficulties which
finally led him to leave his wife. This state of affairs was tolerated
for ten or twelve weeks before the panners asked for D. H. to be
replaced by their regular market-man, G. A. They had had the
opportunity of deciding whether G. A. was a workman fitted for
their job; had they been dissatisfied with him he would have been
rejected, and D. H. would have kept his place until someone turned
up on the market who suited the team. Of course, the management
often initiates the introduction of a trainee into the team, and he
will be rejected only if he is unsuited to the work or to the team,
or if he is thought to be in competition with one of themselves.

If the men feel that no additions to their team are necessary and
perhaps the management is 'gunning' for one of their mates they
will oppose such an introduction.

[1] The market may also contain a very few men who prefer to have their work irregular.
so that they can stay away from work if they wish without such ill-effect as would result
if they had regular contracts.

Day-wage men

'*Bye-workers*'. Men paid at the minimum day-wage (24*s*. 11*d*. in
1953) or just above it are haulage-hands and 'bye-workers'. The
'bye-workers' are sent, usually in pairs, to do odd jobs such as
clearing up unfinished work, removing old timber or rails from
disused gates, or removing the debris and making the roof safer
after a minor fall of stone. Often these 'bye-workers' are old and
partially disabled men who take these 'light jobs' in the period
between the end of their days as contract-men and the beginning
of their retirement. Some men stay on contract work in full health
until retirement, and there are one or two examples of men over
65 working on heavy contracts; but usually a man leaves such work
in his fifties, and, of course, the incidence of ill health and disable-
ment forces many into lighter work long before this. The collier
or any other piece-worker is in his prime between 20 and 40 but it
is unlikely that he will get a regular job on the coal-face before the
age of 25 or 26. Certainly statistics show that the proportion of
face-workers to the total employed was increasing steadily in the
first four years after the war, but the latest figures suggest that this
tendency may have been arrested.[1] If this is so, the result will be
perpetuation of the state of affairs where most of the places on each
newly developed face are taken by men from other faces now
worked out, and young men step into face-work only when an
older worker leaves or dies. This induces a degree of competition
among the young men seeking the best jobs, the jobs at the face.

To return to the day-wage men; the bye-workers certainly do
not work as hard as the contractors, and there are innumerable
jokes about taking it easy on a bye-work job–any man seen sitting
down will be asked if he is bye-working.

[1]Table illustrating proportion of face-workers to total of wage-earning miners
in Great Britain

Year	At the working face	Others underground	On the surface
1943	286,000	260,000	161,300
1945	282,500	267,800	158,600
1949	296,200	262,600	161,100
1952	293,600	261,000	161,000

For the proportion of face-workers at Ashton Colliery see p. 38.

However, these men are experienced, and they may be sent to all manner of jobs, often jobs which inexperienced men could not manage; in some circumstances an overman will ask bye-workers to help out with piece-work, and they are paid accordingly. The most common complaint from bye-workers, like market-men, is that they are 'messed about' because their work is the least planned of all in the pit. They might have to walk several miles in a shift, going first to one job, then to another. Like the market-men they can expect day-to-day variation in their work. Those market-men for whom there is no contract work are sent on this sort of work, for which they are paid the minimum rate plus a small allowance payable to skilled workers.[1] Should a skilled market-man find himself constantly sent on such low-paid work, he will first approach the colliery officials for some sort of promise of a regular job; as a final step he will seek work at another colliery. Generally speaking there is little socially significant difference between the ordinary bye-worker and the collier, for the bye-worker is usually an ex-collier, retaining his characteristics, and because he has not taken up the 'partnership' type of piece-work he has not the individualist and 'collaborationist' tendencies of the well-paid contractors or 'big-hitters'. One must except, of course, those contractors of the latter type who have themselves descended in later life to bye-work.

Haulage-hands. Our final category also consists of day-wage men, the haulage-hands. In the very great majority of cases coal is transported, once it leaves the loader-gate, in tubs (or trucks) to the bottom of the winding shaft. These tubs contain up to one ton of coal (exceptional modern methods use larger trucks) and run on steel rails laid on wooden sleepers. Empty tubs travel on one side of the road, full tubs on the other. A coal-face may be up to three or four miles from the pit-bottom, and only rarely will the roads be straight. Typically the tubs are pulled in runs of twenty to fifty by a steel rope worked from the drum of an engine underground. This rope runs at a speed of not more than three miles an hour and pulls the tubs by means of steel clips or clamps. During coal-working hours, i.e. on those shifts when coal is worked and hauled

[1] The so-called 'skilled-shilling'.

out of the pit, the rope runs continuously except when stopped by
the system of electric signals which exists for the safety of haulage-
workers and the easier carrying out of their work. This signalling
system is constructed so that the rope can only keep moving when
every haulage-hand has signalled that he can work safely. A code
ensures that only a man who stopped the rope can start it again.
The same system has been adopted in the running of conveyor-
belts.

Haulage-workers man the road transporting tubs at intermediate points from conveyor-end to pit-bottom. At the end of each
conveyor works a team of men who divide among themselves the
following series of tasks: working the belt and its loading 'hopper'
as it unloads coal into the tubs, and sending them away on the
rope. The arduousness of these jobs at the loader-end varies a great
deal. In the early stages of development of a face the haulage-hands
may be working with poor roads and lack of permanent junctions
and points. It is difficult to describe the obstacles and frustrations
caused by these factors. The weight of a full tub can easily displace
a badly laid roadway, and lifting a full tub back on to the rails is
'horsework' in the best of conditions. At a bad loader-end this
may happen regularly, and with coal filling from the belt at well
over a ton a minute, a tub off the road can be a cause of great delay.
If the belt cannot be stopped immediately, chaos ensues in that the
whole roadway is blocked with coal and the job cannot be restarted
until it is removed. Scenes like this, except when conditions become
settled and better organized, are by no means uncommon and
together with other breakdowns on the transport system they are
the worst feature of the haulage-hand's work.

Should a run of tubs along the road run into a fall or a break in
the road between junctions then the whole run may be overturned
before anyone sees enough to signal the rope to stop, and there is
half an hour or so's hard work for every available man. Every
miner knows that since 1945 such incidents are very much on a
decline, with wider and higher roadways, efficiently laid rails, and
the introduction of electric lighting on a much wider scale than
before. At one time only the pit-bottom had firmly laid roads, a
thoroughly secure roof, and adequate lighting. This was because

efficiency in the working of the shaft is an absolute prerequisite for a colliery's keeping at work. With the rapid improvement of the roadways the work of haulage-hands takes on more and more a routine character which distinguishes it clearly from the nature of the collier's work. Haulage-work is traditionally the work of 'lads' though there are also always to be found a few men, older or partially disabled, on some sort of haulage work. A typical haulage-worker might have the job of removing the clamps on runs of empty tubs as they round the bend at his junction, and replacing them on the rope once they are safely past. According to the scale of operations, he or his mate will do the same job with full tubs going the other way. Most haulage jobs need fairly constant attention from the beginning of one shift to another, but there may be one or two 'bobbies' jobs' in a pit, such as performing one operation twice an hour; this will depend on how up to date the colliery happens to be.

It is perhaps in haulage work that the greatest change has taken place recently in the work of miners, at least in the West Yorkshire coalfield, where the most modern methods of mechanical cutting and loading are unsuited to natural conditions and have therefore not revolutionized the work at the face. There are still survivals in modern collieries of the older conditions of haulage work. Ponies are still used to haul tubs where it is not worth while or possible to install mechanical haulage, e.g. on development work or in isolated sections of the pit. Thus one man may have the job of 'driving' a pony with twenty empty tubs to a pair of rippers and taking away the full tubs one by one to the main roadway. This was the normal method of work in stone and coal in the pits a score of years ago and more. Development faces will sometimes use old and obsolete engines for hauling the small amount of coal and stone they fill. It is jobs in this kind of work which are far the most arduous for haulage-hands. The survival of such methods serves to emphasize the great changes in the young miners' work.

Day-wage work and team-work

Only a brief comment is needed on the relations between men engaged on day-wage work. Haulage-workers certainly have to work smoothly together in those places where they work in small teams of four or five – at a road junction or at a loader-end – and in the pit-bottom where a large number of haulage-hands must keep the tubs running very quickly, running empty tubs into the pit and loading full ones into the cage at a rapid rate. It is clear that they are not mutually dependent in the same vital sense as a team of colliers. Many accidents do occur on the haulage roads but they are usually neither as serious nor as numerous as those at the face; there is no parallel in haulage work to the great personal responsibility for his own safety held by the collier. But the principal sense in which the colliers, and other groups of contractors to a lesser degree, composed a true team, was in their dependence on the skill and strength of every other team-worker to complete his job and ensure the present and future security of the group: their piece-work is *in the long run* collective piece-work. No such factor exists among day-wage workers,[1] so that the haulage-worker or bye-worker is much less interested personally in the efficient completion of his work. A bad workman brings more work on his workmates, but there is not the same urge to work well as among piece-workers. This difference is given added emphasis by the more monotonous nature of the work of the haulage-hands.

His job compares with that of many factory-workers in that it makes movement dependent in detail on the workings of the haulage machinery, just as an assembly line-worker's movements are regulated by the power plant of the factory.

THE RESIDUE OF PAST HOSTILITY

No topic of conversation is more common among the older men in mining villages than the contrast between the lot of the miners today and in the past, particularly between the wars. The bitter memory of the economic slump and depression is a commonplace for those who talk about miners, for it is known that the miners had experiences more severe than those of almost any other section of the community. But the old miners discuss a great deal

[1] Occasionally the haulage-hands working the conveyor- or loader-end are paid on a piece-work basis.

the actual conditions of work in bygone days, and it is necessary to give this some prominence, because these older miners see the changes as productive of different types of men in the coming generation of miners. Pressed, they will say that fundamentally mining is a rotten job and always will be and miners will always be the same, but they hold strongly to the view that because of the changed industrial and social conditions, the younger miners are in some way 'softer' than they were. Despite the fact that the 'longwall' system of coal-getting, described above, is of long standing, it has only recently become general, and its combination with coal-cutting has replaced a method which required both more skill and strength from the collier. This method was operated either by the individual colliers or in pairs, who worked their own 'stall'. Of course, it was still the case that a number of men would be working near to each other. The essence of the difference between this and the modern system of coal-getting was that the coal was not mechanically cut, explosives were much less extensively used, and the coal was not removed by conveyor. A number of colliers would be supplied with tubs by a driver and his pony. This driver would be a young man–a 'lad'–and in these earlier days no man became a collier who had not gone through a period of driving. In this sense driving was part of every collier's life, and one often hears descriptions of the lot of the driver in the time of the youth of the older colliers of today. The work of the driver today can be awkward enough; his horse may be wayward, obstinate, and even dangerous unless he knows the technique of handling it. Road conditions may give him the backbreaking job of lifting many full tubs back on to the rails. The old colliers remember driving when there were roadways hardly high enough for the horse to pass under, and roads in much worse conditions than any today; all this in addition to the necessity of quickly getting the full tubs away and keeping the colliers satisfied with empties. The apprenticeship was a hard one: the colliers needed to work as hard as they could at their own job to make anything like a living. For years on end they could depend on only two or three days' work a week: thus they were reluctant to spare any thought or effort for the difficulties encountered by their young driver.

When a man graduated from driving to 'coaling' he found himself in a job entailing sweat and concentration for six or seven hours on end, and an incessant struggle against the deputies and other representatives of the management to keep his wage at a decent level. It is very important to recognize this violent antagonism between man and management in all its sharpness as only an apparent contrast with present conditions. The collier will describe it as a change in the balance of power. In the 1920's and 1930's the bosses still held the whip-hand, and they came down heavily on the wages and conditions of the miners; the shortage of miners and the national need for coal in the post-war period have reversed that position. "We've got the whip-hand now and we've got to make it bloody well crack." It is in this sense just as much as in the arduousness of his toil that the lot of the miner has changed.

In the 1920's there still persisted in parts of the West Yorkshire coalfield the 'butty system' – a contractor or butty-man took out a contract at a price agreed with the manager and paid the colliers himself out of the weekly proceeds. The butty-man usually worked at the pit himself, but old colliers complain of butty-men who sometimes appeared only to pay out on a Friday. In their relationship to the management these butty-men were, of course, the extreme of the tendencies observed among some contracting stone-workers today. Their interests rested quite clearly and unequivocally on good relations with the management, and were opposed to those of the colliers. For the most part, however, the butty-system in the 1920's was giving way to the system of direct contracts between colliers and management. But the day-to-day tussle between the collier and his deputy over the wage-note was only part of the general antagonism between miners and coal-owners. This antagonism was no doubt exacerbated by the very hard nature of the collier's work. The following examples are quoted to illustrate the combined effect of all these factors on the collier of yesterday. They illustrate also the opinion among the older men that modern conditions have their effect on modern miners.

"A man of 55, now earning a very good wage as a loader-gate ripper, spoke with great feeling to a young newly arrived workman as they

walked a mile and a half out of the pit together. How easy it was to
walk about in the pit these days; he remembered when you had to
crawl about like an animal. In those times you had to make your mind
up from the very start to take a stand and show your guts. If you showed
the slightest signs of weakness in the face of the overman's blustering
and cursing when you started the job, if you did everything you were
told without demur, then life would soon be hell for you. You had to
make it quite plain that you could swear as well as the next man, give
as good as you got—and if you didn't they'd put on you viciously—'the
bloody swine', he ended with great feeling. The same man often spoke
for long periods about politics to the newest and youngest member of
the ripping team, a lad of 21 who had just finished his training. Typically
he introduced the subject by pointing out that although wages and
employment might be secure now (in 1953) this was a position that
had had to be fought for, and with the existence of a Conservative
government these conditions might again be attacked. It was essential,
he insisted, not to be deluded into complacency by the comparatively
happy position in which they now found themselves. This piece of
instruction came from a man with no special interest in political
matters, and far from being an isolated example, it represents the general
trend of discussion whenever political questions arise in conversation.

"Another man in his late thirties, J. S., spoke in the following way
about politics and work, with very little introduction and with no
prompting. He was a Socialist, he said, and he spent some time in
contrasting men like Attlee and Bevan with Churchill, a 'turncoat' who
would like to turn the troops out against strikers if he thought he could.
However, the time was past, J. S. thought, when such a display of
force by the State was possible. There was a younger generation now,
and they with their independence, self-confidence, and initiative, know
how to look after themselves. However, for this generation working
conditions had always been good, and their lack of experience of times
like the 30's was partly responsible for the return of the Tories in 1951.
When the election results had been announced, he said, he had expressed
the hope that the Tories would squeeze the workers hard, and give
them a chance to learn."

It is worth remarking here that two influences seem to be at
work when men make remarks of this sort, which are, of course,
very common and by no means confined to mining or even to
working-class life. The older workers do certainly feel that the
younger men take their improved standards too much for granted
so that they may tend to neglect the organizational and political
necessities bound up with those improvements. Another element

seems to be a certain degree of resentment or jealousy that young folk are enjoying in the mining areas today a better life than the older miners ever experienced. Somehow this offends their notions of justice and fair reward: after all, every man and woman should be rewarded in a manner commensurate with their effort. The slogan–'a fair day's pay for a fair day's work' is a good example. The young miners have stepped into a situation made more pleasant by the struggles and bitter hardships of their predecessors –and they seem hardly to be grateful. This is how the picture is seen by the older miners. The suggestion that 'a little hardship would do them good' is a characteristic one. It is not suggested here, however, that this 'resentment' is more basic or powerful than the 'political' reason given in our example, for this also is very common. A political reflection of these attitudes is the theory put forward in Labour circles that only a slump or other serious economic crisis can bring any militancy and political awareness from the working class–'they'll never move forward till they get a kick in the backside'.

The old antagonism of management and worker lives on today, in fact is very much alive. In 1953 the North-Eastern Division of the National Coal Board announced the formation of efficiency committees at all pits in the area for the increase in productivity and to smooth out difficulties. The following is a verbatim report of a conversation between colliers from two collieries in the No. 8 area of the division:

> "S. F. 'I was at the Union meeting and they were saying there's five on the committee, two representing the men, two from deputies and the shotfirers, and one from the administrative side, that'll be managers and undermanager.'
> "J. C. 'Didn't you tell them that's three to two?'
> "S. F. 'Yes, I got up and said we're beat before we start.' "

Our final example is taken from conversation with A. B., a 65-year-old native of Ashton who for twenty years has been unfit for work, so that his memories of work are strongest for the period of 1920–33.

"He began by commenting on what he thought an uninformed discussion on the B.B.C. in a programme on whether nationalization is or is not an improvement. This brought him immediately to the description of how bad pit-work used to be.

"There had always been deliberately cultivated a group of 'good men' or 'bosses' men'. These the boss would maintain were good workers, but they were 'well in' because they could be relied upon to put good relations between themselves and the boss before principle. A. B. describes how when such a man was sent with him to work (he claimed that he was a very good collier, sharing the record with three other men for the amount of hand-go coal filled in one shift) he would sweat the life out of him by working at a tremendous rate–'and some of 'em weren't strong enough to bust a bladder'."

"The management, he said, were continually double-dealing. He gave the following example. The most difficult part of hand-getting is breaking in and making a face ready; at this stage the coal is at its hardest and more skill and strength are required than at any later time. After the first week or two the work becomes normal. A deputy sent A. B. into one of these 'holes' with two other men, their third mate being off work. . . . A. went, only on the understanding that he would be a regular collier in that place, and helped in this hard, preliminary phase. When the other man returned, A. went to see his deputy, now a different one, and appealed for his job without success. 'Them's the sort o' tricks they used to work on you.' He described how if men in one 'hole' were earning high wages, the bosses would send in day-wage men when they knew that the work was at a stage when no extra help was required. Such men must be paid 'off the colliers' note', and this decreased the piece-worker's earnings. Another device advantageous to the boss was the payment of work in advance. Men often had to ask, say, for a yard of ripping to be recorded in advance to ease financial circumstances. This disqualified them from another claim which could be made on completion of the work. 'The collier always had hold of the dirty end of the stick.' When A. B. had been sent as a day-worker to such jobs where there was no extra work, he would ask the deputy for a note to go out of the pit. In this he was exceptional. Later in this chapter it is made clear that this fear of being sent to another man's job persists in present-day industrial relations in mining.

"A. B. spoke too of the wages of miners now as compared with twenty-five or thirty years ago, adding that 'in them days coal were got with *that*'–pointing to his shoulder and arm muscles. This 'better money' he saw as a consequence of nationalization and Labour's rule, just as he did the stronger position of the men *vis-à-vis* the management.

"He went on to criticize those Tories who blamed the Labour government for our bad economic position. 'Take unemployment,' he said, 'and history proves that it was Churchill and the Tories who started the lot.' After the 1926 strike the miners lost the 7-hour day;

thus greater production was possible from the same number of men; in this way the number of out-of-work miners had been increased. Unemployment had now vanished; so how could miners not support nationalization and the Labour Party? He was disgusted with 'the idiots on the radio', and wished he could get on the air and put across a few home truths.

"Though they were now old men, there was still hostility between A. B. and some of the old 'good men' ('arse-creepers'), and he commented on a rumour that one of them had said he was lazy, in this manner: 'Let him come out and dare say it in front of me, he daren't' – adding an oath.

"The same man (A. B.) in a discussion over large volumes of ale on Christmas Eve crossed swords with his two sons-in-law, both contract-workers under 30 years of age. The younger man regarded those who were elected into local union posts as grafters and place-seekers. These local officials drink together in the same clubs and arrange union business so that no outsider can ever take the initiative. Arthur B. took the line that he'd known many leaders to go the wrong way but it was up to them to keep fighting, to go to their meetings and speak up, otherwise nothing would ever be done. The young men violently protested that this would be utterly useless for they could never find leaders who were not changed men after their election. A. B. here fully exemplified the old militancy as compared with the complacency of the younger group; however, it should be said that a great many, probably a majority, of the older men, are just as cynical about organization and officials as were A. B.'s opponents in this discussion.[1]

"On another occasion A. B. returned to his condemnation of the coal-owners for their deliberate attacks on the miners after they had defeated the 1926 General Strike and defeated the 7-hour day. The same production level was maintained, while men were sacked, leaving only one or two bye-workers on each shift. The result was that although coal production did not decrease, transport and road conditions deteriorated considerably and the getting of coal out of the pit depended more and more on men and horses 'pulling their guts out' with roads seldom cleared and bad places only rarely improved. This policy left the pits in shocking condition for the advent of nationalization, and all-in-all bequeathed a legacy of bitterness among the men, who for so long had to work in conditions where their comfort was the last consideration. A. B.'s last remarks are borne out by any observation of labour relations since the nationalization of the mines."

Such reminiscences of bitter relations with management are extremely common. Miners will relate with pride occasions when

[1] See Chapter III, 'Trade Unionism in Ashton'.

despite the severest economic difficulties they had acted according to strict principles. An old trade unionist, E. H., told of how he and a group of Ashton men took jobs, after many months out of work, at a colliery some twelve miles distant. Even with employment as insecure as it was at the time (1933) this group has steadfastly resisted any encroachment on their conditions of work. One day the deputy approached one of their group and deliberately provoked trouble.

"You'll be working this seam for 2*d*. a tub less from today."

"There'll have to be a proper discussion with our representative," came the reply.

"If you don't want that rate of pay, you can go home," said the deputy, and he maintained the same attitude when confronted with E. H., the 'checker' or chargeman, i.e. the spokesman of the men. At E. H.'s word every man put on his coat when the deputy remained inflexible, walked off the job and walked back to Ashton at dead of night rather than sacrifice principle because of severe economic difficulties. E. H. contrasted this type of event with trade unionism in mining today, readily admitting, however, that the causes of such actions were more rare nowadays.

CONFLICT IN THE INDUSTRY TODAY

In the present period there seems to exist something of a contrast with all this. Since 1926 there has been no nation-wide strike sanctioned by the National Union of Mineworkers, and since 1939 every single dispute has been unofficial, unsupported by the National Union of Mineworkers. In Yorkshire the famous Grimethorpe dispute of 1947, in which the men were finally proved right and which provoked sympathy strikes all over South Yorkshire, is the best remembered example of a common tendency for the National Union of Mineworkers leadership to join with the National Coal Board, the Government, and the Press in urging the miners back to work. In the Winders' Strike of December 1952, the National Union of Mineworkers actually put itself in the position of promising to help supply substitute winders to those pits where winders were on strike. Events such as these suggest that the apparent calm in relations between labour and management in coalmining is something of a surface phenomenon. The negotiating machinery developed during the 1939–45 war, the fear of repetition of 1926, and the definite improvements brought with nationalization have all contributed to an absence of open and organized struggle between the National Coal Board and the National Union of Mineworkers at national level; but certain processes inside the industry are becoming more and more difficult to conceal. It is true to say, for example, that scarcely a day goes by without at least one pit in Yorkshire experiencing a strike.[1] The miners of Ashton, in common with many others, have spent months in haggling disputes with the management over wages and conditions, with a resulting severe loss of work and income. In these disputes (to be treated more fully in the next chapter) the heritage of suspicion and bitterness derived from earlier days was much in evidence. But to say this–as so many do–is not enough. Far from disappearing, the attitudes of suspicion flourish. Men who are ordered to do a particular job when they are 'on the market' are quick to suspect that they might be sent to cut down the work of certain contractors, and they will often say 'I'm not going to do somebody else's work, am I?' and a union man will be standing near to gain assurance that the management does not get the chance of splitting the men in this way. One of the authors was sent one day during a short working period at Ashton Colliery to help two rippers. The men refused to start work until they had seen the deputy and made sure that their wages would not be affected.

CONTRACT WORKERS AND WAGES

In discussion with officials of the National Coal Board an attempt was made to discover the attitude of the administration to the large number of disputes in the Yorkshire coalfield. One principle dominated the replies: 'Every strike is a wages strike.' Men may go on strike because of bad roof conditions, because of water, because of difficulties caused by mechanical breakdowns; but in all these cases it is the effect on the pay-note that is really at stake. It is not entirely true that the miner is concerned solely to get the maximum amount of money for every single task he

[1] There are 116 collieries in the area.

performs. The wage structure, with its differentials for skill and danger, has been built up in such a way that a man's status as a strong and skilled worker, and as a man worth his pay, is conveyed by what shows on the pay-note. Any factor helping to cut down the amount of his wage is an attack on the whole status of the contract-worker. All this adds intensity to his attitude towards wages questions, an attitude which follows logically from our remarks on the position of the wage-labourer and his relation to his work. The fundamental relation between worker and employer is most concretely expressed in the division of the product, in the size of the worker's wage. The method of piece-work and fixing of contractors' wages in mining is only a very clear example of this principle.

When a 'price-list' is agreed between a collier or group of colliers and the management, it is known on both sides that all the work involved cannot be fixed at a definite rate per day. Conditions of safety and efficiency vary, and allowances of all sorts are included in the price-list. It is in the claims for these allowances that the daily opposition between men and management is manifested. Conditions may require the erection of extra supports or the completion of certain jobs which have been hindered in the previous shift. The collier constantly presses the deputy for the inclusion of such tasks in the record of work so that he will be paid extra for them; the deputy tends to try to work his coal-face cheaply and avoid such payments where he can. A price-list may include agreed payments, say, for the existence of water at the face when in sufficient quantity to interfere with the work, but there will inevitably be a more or less prolonged wrangle as to whether the water present is sufficient to warrant the payment.

Piece-work of this kind is paralleled to some extent in other industries, but the direct confrontation of the miner with nature brings an element of unpredictability which is unique. Men on piece-work in an engineering factory may perform the same operation the same number of times each day for many years, and their productivity can only be affected by a breakdown of the power-plant, of the individual machine, or in the smooth running of another department. In any of these events, his remedy is usually quite simple; he 'books off the job' and is paid in the interval on time rate, so that his piece-work earnings are unaffected. The ease with which he is able to do this depends on the strength of trade union organization in the shop or factory. The piece-worker in mining experiences a similar type of breakdown in addition to the 'natural' events which may affect his wages. Mechanical breakdowns of any kind, which lead to the impossibility of keeping the face conveyor on the move, are made up for by 'waiting-time'. A team of colliers is paid its total resulting wage on the 'big-note' – issued to the team on the day before pay-day. This is then divided by the number of man-hours worked by the team, and each man is paid for his hours worked at the resulting hourly rate. Now it is obvious that if some of the time the belt is not moving and coal cannot be filled, the rate per hour is going to be decreased. There is, therefore, usually an agreement to pay so much per hour for any waiting-time, i.e. time spent waiting for the belt to start working, over and above the wage paid on the price-list. Problems of whether waiting-time should be allowed or not cause just as many disputes as the other factors interfering with production. The collier cannot 'clock-off' his job and 'clock-on' to waiting-time as can the factory worker; he must make representations to the deputy to have the waiting-time recorded. Overtime is the subject of similar disputes; the factory worker 'clocks-out' whenever he finished work, and there can be no argument. Whether a miner is paid for his overtime depends on whether the deputy enters that overtime in the daily record.

This position of the deputy in the wage conflict, i.e. as the representative of management towards the men, makes it extremely difficult, one might say impossible, for him to be the safety official visualized in the 1911 Act, or the efficiency promoter apparently desired by the National Coal Board. It is this responsibility for wages which has led to the position where deputies have always been 'bosses' men' and since the cause of this has not been removed, they are still regarded as such. In the conditions of the early days of nationalization, many deputies, especially of the younger generation, took a generous attitude towards allowances and overtime, and the men recognized this, but the emphasis on profitability by the National Coal Board, the constant campaign for economy, and 'cutting costs', is naturally not conducive to such an attitude from deputies. The groups of men waiting at the pithead on a Thursday or Friday are evidence of the wastage and backwardness of this system of wages, and there is no wonder that colliers see little change in this respect from pre-war days. Any man or group of men dissatisfied with their wage must wait for the deputy or other official to come out of the pit before any detail can be challenged; men have often been known to wait the whole of Friday afternoon and an hour or two on Saturday morning to see the deputy or the manager. A manager at a neighbouring colliery to Ashton had the reputation, which his actions did nothing to belie, of staying down the pit longer on Friday than on any other day in order to avoid facing those men who disagreed with their wage-note.

Two general observations are relevant to this discussion of piecework. Firstly it is clear to the men themselves and to any outsider that the skill of the trade union representative of the miners in negotiating price-lists and arguing with the officials will determine the ease with which the collier gets his due in the wage-packet. If these representatives are weak, tongue-tied, and slow-witted they will lose all along the line; the men will therefore constantly be involved in petty disputes and prone to engage in strikes or delaying action against unsatisfactory conditions. It seems likely then that efficiency will be greatest in those pits where the trade union representatives are most skilful in negotiating for price-lists, and where the management recognizes this. Secondly it is suggested that the very fact of piece-work confirms our remarks about the status of the wage-worker. Speaking 'logically' and assuming that every worker was as interested in doing the job well and achieving maximum production as the management is, the best results should be obtained by paying each collier a good daily wage and asking him to get on with the job. Piece-work, in mining at any rate, is not just an 'added incentive'. It is the natural form for the wages system to take where there is a tradition of hostility, and in mining it takes on its greatest impetus and strength from the physical nature of the work, as has been shown. Significantly

enough, no alteration in this wages system has accompanied
nationalization; payment by results does not necessarily imply
hostility between management and men, but it should have been
recognized that the practices revolving around piece-work in
mining have, in fact, been built up on such a basis of hostility, and
this atmosphere still surrounds piece-working. A change would
have been difficult, since a powerful motif in the miners' view of
industrial relations is 'what we have we hold', and any attempt to
affect a change in the payment system would have been greeted
with distrust.

The existing price-lists have been fought for through years of
dispute with the management, and any attempt to throw them
overboard would have been interpreted as a ruse to lower the
miners' standards.

DAY-WAGE MEN AND WAGES

The day-wage men do not confront the management in their
constant wages struggle in the same way as the piece-workers.
Their daily wage is paid regardless of the work they do. Only
overtime-working can supplement their income, and it is mainly
on this account that they will have slight brushes with the management. From this it would certainly not be safe to assume that the
day-wage men are the most 'satisfied' of the mineworkers. Face-
workers are in a better position and have greater opportunities for
fighting on the wage issue than do haulage-hands and surface-
workers, and this is the main reason why most disputes occur
among face-workers. An additional reason is the higher level of
wages in this section, which makes the worker more independent
and secure. Throughout the country since 1945 the miners have
been thought to enjoy abnormally high wages; it cannot be too
strongly stated that only contract-workers are highly paid, and by
no means all of these are above the £10 per week level for a
normal five-day week. It is sufficient to state the normal rate for
underground day-wage workers and for surface-workers to see
that their wages compare by no means favourably with those of
other industrial workers. The minimum (in effect, normal) day-
wage for underground mineworkers over 21 is 24s. 11d.; for

surface workers it is 21s. 3d.[1] The Press seems to be very fond of quoting 'average weekly earnings' which amount throughout the industry as a whole to just over £11 weekly! This figure is extremely misleading. Among other inadequacies, it veils the fact that a large section of mineworkers is working at poor rates of wages, a fact which leads to a simmering discontent and dissatisfaction. Only very slight increases have been effected in the minimum day-wage over the past five years, and many miners, particularly surface-workers, work very long hours overtime–often making a total shift of 12½ hours–in order to supplement their wages.

The face-workers can fight for higher wages at colliery level by direct negotiations with the manager; they can more or less assure themselves of a 'good week' whenever they have a mind, as is witnessed by the record output and attendance at 'bull weeks'.[2] But the men working at a national minimum, apart from overtime, can only see a change in their wage as a result of national negotiations between the National Coal Board and the National Union of Mineworkers. The National Union of Mineworkers is certainly powerful enough to win any demand for a day-wage increase; of that there can be no reasonable doubt, having regard to the country's need for coal; and yet the union has not used its real strength to this end. No doubt one very powerful factor contributing to this tardiness on the part of the union is the top-heavy structure of mining trade unionism. This feature arises from the quality of the issues tackled by local union branches and at area level over the years. The strength of the organization was built up when the normal development was for nearly every man, after a period of haulage work, to graduate to coal-getting or other contract work. The mature men, those more likely to take part in union activity, therefore had a common interest; and since every young man expected to reach the position of face-worker eventually, a union leadership composed of face-workers, and more particularly colliers, was suited to the needs of the men, and arose naturally from the conditions of mining. Later developments in

[1] Figures for June 1953.
[2] The week's work on which wages prior to a holiday will be based.

mining led to the need for a larger proportion of day-wage men. Their conditions, as suggested above, do not bring them into constant disputes like the face-workers, so that no change in trade union structure accompanied the changes in industrial organization. Consequently the branches and higher levels of the National Union of Mineworkers are dominated in the leadership by contract-workers who not unnaturally do not have the problems of day-wage men so much at heart as those problems with which they are more familiar and are even personally concerned. The National Union of Mineworkers in this sense is top-heavy, and the position in the local colliery branches makes it difficult to change the situation, for here also it is among the face-workers that good negotiators and militant representatives tend to be found. These local organizations each send a representative to the Yorkshire Area Council at Barnsley. Thus neither at branch nor area level is there much possibility of a decisive initiative on the issue of the day-wage, any more than at the level of national negotiations.

In the Ashton collieries this division between poorly paid day-wage men and better-paid contractors has not been absent in recent disputes. When a coal-face dispute leads to strike action there is simply no work for haulage-hands and they lose their wages. This is sharply appreciated by the day-wage men; and they are not slow to compare the situation with the failure of the face-workers to show any real interest in the fate of the minimum wage. In the summer of 1953, in the Silkstone seam of Ashton Colliery, a prolonged series of short stoppages culminated in a stretch of several weeks during which only a few days were worked. It was among the wives of the miners, towards the end of the dispute, that complaints like this were often heard: "My husband is a day-wage man, and he has no work because these contract-men are on strike; and they already get twice as much as him, some of them!"

SPECIAL FEATURES IN WORKER-MANAGEMENT RELATIONS IN MINING

In one respect miners working at all levels of wages have a common grievance. When their leaders signed the Five-day week Agreement in 1947, the old level of wages paid on six days was only to be paid for five shifts if all five shifts of the week were worked. In other industries the level of wages had been maintained, on the winning of the five-day week, by raising the hourly or daily rate. The miner gains from the agreement only if he works his full five shifts, i.e. the miner, if he works four days instead of five, loses two days' pay. It is, therefore, not surprising that some miners will prefer to lose three days' pay for two days, or to 'save' their absenteeism. In any case, it seems unkind that of all industries the one characterized by most sickness, injury, and fatigue, should be saddled with this clause in the hours agreement. The National Union of Mineworkers is now pressing for the division of the weekly bonus, i.e. the sixth day's pay which accrued from five days' full work, into five daily parts. Another worthwhile note to any statements about the level of miners' wages is the fact that in no other industry is so much working-time lost through the agencies already mentioned of injury, fatigue, and illness. The paucity of welfare conditions and the great danger of accidents in the mines as compared with factories, exacerbate this state of affairs and add to the unattractiveness of the industry. Mining is governed by the Coal Mines Act of 1911, though another Bill is now before Parliament (1954).[1] It is instructive to compare certain points in these documents with the Factories Act, as part of a general summary of conditions in mining as compared with industry in general.

The proceedings following the Knockshinnoch disaster in 1952 showed the 1911 Act to be particularly weak in that it is very difficult to apportion criminal liability for loss of life and limb. The projected Bill (1954) lays down that no criminal liability exists (a) if 'due diligence' was shown by the officials concerned (Sec. 136) or (b) if the 'act of omission' constituting the alleged contravention was due to an act of nature which could not reasonably have been expected to be provided against (Sec. 137). Many commentators have pointed out that this is a 'loophole' clause which has no parallel in the Factories Act: similarly there is nothing in the Factories Act akin to the absolution of a mine-manager who

[1] Since this study was made, the new Coal Mines Act has become law.

can prove ignorance of the omission of the alleged offence. The safety code laid down in the new Bill is only a little more firm than that of 1911. In coalmining it would be reasonable to expect definite and firm standards of safety, and yet the Bill constantly uses the 'reasonably practicable' clause. It is difficult to see how a satisfactory safety code can include such provisos in an industry like coalmining.

In another respect the new Bill will remedy the existing inequality between the miners and other workers. The 1911 Act had set limits on the ability of the men to sue for damages from industrial injury. This clause now disappears and reasonable compensation is assured. No doubt this provision will affect the safety policy of the National Coal Board. Certain provisions for welfare are made in the Bill. Many collieries in the last fifteen years have been supplied with modern pithead baths, canteen and sanitary conveniences on the surface, and the new Bill gives power to order such extension of these facilities "as appear to the Minister to be expedient for the purpose of securing the 'welfare' of the workmen". Many collieries, including Ashton, are without baths, and at the Ashton northern shaft where 500 men ride in and out of the pit daily, there are neither washing facilities nor lavatories and the 'canteen' is a tea-hut open only at two periods of the day. Underground welfare conditions are shocking. Electric lighting is confined to certain parts of the roadways. Those gates which have roof lighting are showpieces in Ashton and surrounding collieries. In a highly dangerous industry like mining it is fantastic that provisions for adequate lighting should not be stipulated by law as they are in the Factories Act. We have already described 'snap-time'–a short break of twenty minutes for refreshment. No drinking water is provided underground–men carry water in bottles–and there are no washing facilities so that every man is obliged to consume a certain amount of coal dust at best, grease and oil at the worst, with his 'snap'. Lavatories are to all intents and purposes non-existent. All these conditions must be changed before the miner can really feel a complete change in his status in the working class.

MAN AND JOB

The miner does a job in conditions which are still worse than those in any other British industry, though he can see that improvements are rapidly being made. Higher wages were the first step, and the miner sees this as a sign of his emergence from the lowest ranks of society. In addition, he is proud, as he ever was, that he does a difficult, arduous, and dangerous job which deserves greater appreciation than it ever gets. Miners constantly say that no non-miner can appreciate the nature of pitwork and few would challenge them. When they hear complaints of miners' high wages, they confidently offer an exchange of jobs, and this is enough for most public-house politicians. In the pit itself, among his workmates, the miner is proud of doing his job as a good man should, and to a great extent a man becomes identified with his particular job. Recognition that one job belongs to one man is recognition of that man's fitness and his control of that job. Men in Ashton will half-jokingly say when they have spent a shift deputizing for another man, "I've been Joe Hill (or whoever it is) today." On the morning after the retirement of a 65-year-old deputy, W. H., two men greeted his successor (whom they knew well) "Is tha' Bill H. today then?"

Pride in work is a very important part of the miner's life. Old men delight in stories of their strength and skill in youth. A publican or a bookmaker will often joke about the number of tons 'filled off' each day in his establishment by the old men. Older men in the pit who go on to light work will confide that they can still 'go as well as the young 'uns' but they think they deserve a rest. Men of over sixty still working heavy contracts are visibly proud of themselves and resent any preferential treatment. Another influence may be discerned in this pride of miners in their work. For long, and they know this, mineworking has been looked down on; this is felt strongly, and a man's assertion of pride in being a miner is often partly an attempted self-assurance that he does not care what non-miners think of him.

The identification of a man with his work is reinforced, and reaches a higher level of social significance, by the impact of class relations on the carrying out of work. It has been remarked that one of the tactics of which the employers were suspected by the miners in earlier days was an attempt to provoke competition between workers as a safeguard against the growing of their solidarity and strength. This suspicion is by no means dead, and it recurs in situations where a workman is sent to replace another, or to do any job which is not his own. Before a market-man proceeds to his allotted task for the shift he will ask, "Am I being sent to do somebody else's work?" He wonders whether there will be too many men on the job to ensure a good rate for the regular contractors, or he may suspect that the fact of a face not being filled-off is the result of a dispute, so that in effect he would be blacklegging. These suspicions lend strength to the identification of a man with work commensurate with his skill and status.

It is clear then that the work a miner does and the wage he receives both express concretely his status as a man and as a member of his profession. The changes in mining technique which we have generally outlined have served to bring into the open, in a single example, the validity of this analysis. Some twenty-five or thirty years ago it was still common for young men or 'lads' to spend several years on the less skilled and lower-paid jobs such as pony-driving, and after this period they normally became colliers or other contract-workers, with a considerable increase in wages. The increased proportion of face-workers in mining today has done away with this order of things; young men have many opportunities for training, and not only in contract work but also in mechanical and electrical engineering; their term of service on haulage work is ordinarily much shorter than it has ever been. In Ashton at least this has led, together with other obvious considerations, to a policy of sending the largest possible number of the younger and fitter men to the coal-face, and filling in the gaps in haulage-staff from the older men who have finished with contract work. Now it is not too easy for a contract-worker to admit the passing of his full powers and accept bye-work; but to be drafted to haulage work, 'lad's work', intensifies the difficulty of the transition. The older worker feels humiliated at relegation to the work of a boy.

In every way the example of N. D., a 63-year-old pitman who had just left contract work permanently for the 'market', is typical. For three years he worked with a team of panners. On two successive night shifts N. D. was sent to drive a pony to a pair of rippers; the job involved hauling twenty empty and twenty full tubs of stone from the ripping edge to the main road – a distance of some 200 yards including a right-angle bend and a steep slope. On the same two nights a very young man, A. P., who had only just completed his official face-work training, was sent to replace an absentee from the panners' team of which N. D. had been a member. On the first night N. D. protested that he was not fitted for pony-driving; he was getting on in years and could not move as smartly as a driver should: his eyes were failing and he could not safely locker the tubs.[1] For the whole shift he grumbled about the work, declaring that he would never do the same job again. On the following night the deputy ordered him to the same job but added insult to injury by sending him back to the pit-bottom (about 1½ miles) to fetch a pony. N. D. repeated his protests of the previous night and towards the end of the shift had worked himself up into a thoroughly vile temper. A little 'kidding' and 'egging on' by other men was sufficient to send him into the deputy's cabin at the end of the shift and in the ensuing altercation there emerged the real burden of his discontent.

"What do you think you're doing? *I'm not signed on as a driver; I'm signed on as a collier*, and that's the kind of bloody job I should be doing, not a lad's work. There's that A. P. working on the face, and he's never driven a pony in his life – never done a shift driving and he's on the face."

All this was shouted at the deputy with a number of other personal grievances; it was shouted loud enough for all the men near by, dressing to leave work, to hear. N. D. came away flushed and very pleased with himself, for he had, with this demonstration, removed, or so he thought, any reflection of inferiority cast upon him by his work of the last two shifts.

[1] 'Lockers' – stout pieces of wood or steel which are placed between the spokes of the tub's moving wheels to brake it. Without training and agility this can be dangerous and causes not a few accidents and breakdowns.

SUMMARY

Our last anecdote draws together many of the strands which are woven into the complex pattern of social relations in at least one pit in the nationalized coal industry of 1953. In any of the day-to-day lives of Ashton, or of any other mineworkers, one could discern the interplay of the factors we have described. Many years of hard toil and social conflict had given rise to a social structure and an ideology in mining which were fraught with dissensions, contradictions, and suspicions. The ideology of the days of private ownership, the days of depression, unemployment and bitter social strife, certainly is operative in everyday social relations in mining. It is more difficult to say whether those relations themselves have changed, and this problem cannot be fully treated until some analysis of trade unionism is put forward.

In his everyday work the miner has seen great improvement in the physical conditions of labour; the reward for his labour has been comparatively great since 1939; mining offers complete security of employment in the West Yorkshire and most other coalfields. Nationalization, a long-standing aim of the miners, has been achieved. The prestige of the miner in the working class is higher than it has ever been, and the miner knows this. Does all this mean that the miner has experienced a basic change in his status and role in the society, a change which goes with a transformation of the relation between the miner and his work? In fact no such basic change has occurred. In the first place the actual changes have been absorbed into the miners' traditional ideology rather than transformed it. Secondly, changes within the mining industry, and the quantitative improvement of the miners' position in relation to other workers, have been unaccompanied by any profound modifications in the general economic framework of which mining is a part, or of the social structure within which miners exist. Most miners know, for example, that the first charge on the industry's profits is compensation to the old colliery companies.[1] They know that representatives of those companies were

---

[1] In the first seven years of nationalization, despite a loss in the first year, the industry made an overall profit of £90,500,000. However, the compensation bill for the same period was £103,900,000. There was therefore an accumulated deficit of £13,400,000.

among the many non-workers appointed to the executive and administrative staff of the nationalized industry.[1] They saw no change in the local management of the mines when nationalization took place. In all these ways they see themselves opposed to the same forces as before nationalization. When they are told not to strike because of impeding the national effort, when they hear of economy drives and efficiency teams, they see no reason why they should regard such admonitions any differently from the pre-nationalization period.

What of the other characteristics of wage-working, particularly in the mining industry? Joint consultation, meant to be a method of drawing workers into the management of the collieries, is a failure, if only because the vast majority of the men are not drawn into it, so that their relation to the direction of work has not changed one iota. In the working situation the deputy, who has often been working in the same pit for forty years, still has an identical position *vis-à-vis* the men; to get an improvement in wages the contract-workers must contest views with the deputy. If a miner wants to question his wage-packet he still must go through the tiresome and often humiliating procedure of chasing the manager or the other officials outside of working hours. All these things are not just a matter of "old attitudes in face of new facts", as we so

---

[1] The membership of the National Coal Board in 1953 was as follows:

Chairman: Sir H. Houldsworth; *Deputy chairmen:* Sir. W. Drummond, Sir Eric Coates. The previous chairman, Lord Hyndley, was a prominent coal-owner.

Board members:

Ald. W. Bayliss, ex-miner and ex-N.U.M. official (part-time member).

Sir Arthur Bryan, formerly professor of mining, Chief Inspector of Mines, etc.

Sir Charles Ellis, F.R.S., previously professor of Physics. From 1943 Scientific Adviser to the Army Council.

W. H. Sales, worked underground for six years. Degree in Economics 1920. Teaching post till 1947, when appointed Deputy Labour Director, East Midland Division, National Coal Board.

Sir Geoffrey Heyworth, present chairman of Unilever, Ltd.

Gavin Martin, General Secretary of the Confederation of Shipbuilding and Engineering Trade Unions.

Sir Geoffrey Vickers, V.C., partner in Messrs. Slaughter and May.

1942. Deputy Director-General Economic Advisory Branch, Foreign Office.

1942. Member of London Passenger Transport Board.

1946-8. Legal Adviser to the National Coal Board.

J. H. Hambro, Director of Hambro Trust Ltd., Hambro's Investment Trust, Mid-European Corporation Ltd., and Chairman of the Bentworth Trust, Ltd. (*Colliery Year Book*, 1953).

For an analysis of the membership of Nationalized Industry Boards, see *Men on the Boards* Acton Society Trust, London, 1951.

often hear, but in reality the persistence of very concrete aspects of the old social relations of the mineworker in his industry. Most important of all, as was pointed out earlier, all improvements are seen as the result of fighting by the trade union and the Labour Party, as gains only to be preserved by maintenance of the 'whip-hand'. For the miner, his improved position is a reversal of the initiative within a continuous frame-work of conflict, not an end of that contradictory framework.

The truth of this conclusion is best illustrated by the wage disputes developing recently in the Yorkshire coalfield. In one colliery after another price-lists are being challenged, and a move-ment is afoot to rescind the agreement between the Yorkshire National Union of Mineworkers and the North-Eastern Division of the National Coal Board for an upper limit of 2s. 6d. per ton on price-lists in West Yorkshire. From being an 'unofficial' demand, this has become the policy for which the Yorkshire Area National Union of Mineworkers is now pressing. Alwen Machen, Presi-dent of the Yorkshire Area of the National Union of Mineworkers, in a speech at the Wakefield Miners' Rally in June 1954, first con-demned unofficial strikes, but went on to speak vigorously of the need for increasing tonnage rates. So long as wages were improv-ing in mining, just so long could the union openly side with the productivity drive of the Government and the National Coal Board against any agitation among their members, and just so long did it seem reasonable to conclude that a revolutionary change had come to coalmining with nationalization. Recent developments, as illustrated above, and as confirmed by the observation of the miner's attitude to his work, suggest strongly that the biggest changes must still be in the future.

Within the trade union movement among miners there have been, at one time or another, proposals for workers' control of the mining industry. However, it is true to say that the great majority of the men in the pits, although overwhelmingly in support of nationalization both before and after 1946, had and have no set of definite ideas for the actual changes which they would like to see in the industry. In the main the miner measures his progress by security of employment and ability to earn a good wage. For the last fifteen years, since 1939, these aims have been achieved within the old framework of negotiations and relations between men and management, and they have been accompanied by great improvements in working conditions since the advent of the National Coal Board. There results a narrow outlook on the ability of their representatives and themselves to secure better and better rates of pay from the management, and it is on these lines that the National Union of Mineworkers nationally and at lower levels is campaigning. The result is that the old system of industrial relations, characterized by a fundamental conflict between management and men, is kept healthily functioning. It functions all the better on the basic physical fact of coal-getting which, as we have seen, gives occasions every day for conflict between worker and management.

ASHTON

The fact of common residence is naturally of more significance in a town of Ashton's size than for larger industrial communities. A man's workmates are known to him in a manifold series of activities and contracts, and often have shared the same upbringing.

The effect of a common set of persisting social relations, shared over a life-time by men working in the same industry and in the same collieries, is a very powerful one. In the main, this factor is responsible for the reinforcement and reaffirmation of those social bonds which have been shown to be characteristic of present-day mineworking. Solidarity, despite the division into interest groups among the miners in a given pit, is a very strongly developed characteristic of social relations in mining; it is a characteristic engendered by the nature and organization of coalmining: it is a characteristic that has been given added strength as a result of the high degree of integration in mining villages. A miner's first loyalty is to his 'mates'. To break this code can have serious consequences in any industry, but for a miner his whole life, not only his work, can be affected by the actions and words of his fellows. The 'blackleg' miner must be made a social outcast in every way, and not only at work. This is possible in any situation where the workers in an enterprise are living together in one community and form the majority of that community; naturally it does not apply only to coalminers.

Other influences work towards the conditioning of each miner along the lines of the general values and characteristics of the rest of his community. Ashton like many other villages suffered long-term periods of poverty and unemployment in the years between the two world wars. In the strike of 1926 the men of the village could see themselves as one unified force, facing a common enemy and a common fate. Every miner's wife knew that the struggle to keep her family alive through the seven months of the aftermath of the General Strike was visibly shared by every other miner's wife. The near-absence of any alternative employment made it plain to all the people of Ashton that there was in fact little to choose between them. Not only did the many years of hardship following 1926 create a bond of suffering but the period functioned as a leveller of status differences. In those years nobody was well off, and it is significant that in the 1950's one often hears the phrase, when miners or their wives are discussing some individual who has 'got on' – "I don't know why he should think so much about himself – his father was only a collier same as anybody else. . . ." There will follow a detailed description of where the family lived and of experiences shared with them. This fund of shared experiences ran concurrently in time with the industrial struggles of the miners of Ashton and elsewhere, and the two tendencies have firmly reinforced each other.

The years of depression were not so severe in their effects on Ashton and the rest of the Yorkshire coalfield as they were elsewhere,[1] and so it might be claimed that the solidarity resulting therefrom is shallower than for areas such as Durham, which were so much harder hit. Yet there is a sense in which the particular features of the depression for the Yorkshire area have enhanced the solidarity of communities like Ashton. It was because it served an internal and nearby industrial market, that the West Yorkshire coalfield was not so hard hit as the villages of Durham, with no

[1] See Chapter I, 'Place and People'.

such market and depending on export. But the proximity of towns with other industries, with workers usually not so badly off and under-employed as the miners, served to emphasize the unfortunate lot of the mining communities in West Yorkshire, and give yet another strengthening factor in the integration and solidarity of a village like Ashton. The mining villages themselves, and Ashton is certainly a good example, are among the ugliest and most unattractive places to live; they are dirty, concentrated untidily around the colliery and its waste-heaps, and lack the social and cultural facilities of nearby towns. Passengers on buses going through Ashton will invariably comment on its drabness, and the place is often quoted as an example of the backwardness of the mining areas. In conversation with strangers men and women of Ashton will defend their town almost before it comes under attack on such grounds, for they are very well aware of the kind of reputation it possesses. In the minds of the miners and their families, this has been an added factor in depressing their status within the working class and the society in general. Similarly they have shared together those improvements in mineworking and living standards consequent on the war of 1939-45 and the nationalization of the mines.

The conditions of mining itself in Ashton present little that is exceptional, although particular features of these conditions have helped to give consistency to the picture we have already described. Ashton miners work the same seams, at similar depths and in similar conditions of work as do other miners in the coalfield. However, it is important that in 1935 Manton Colliery, employing 600 men of the village, was permanently closed because it was uneconomical. Again an experience common to many brought a train of difficulties shared by a large section of the community. For Ashton the closing down of a principal source of employment and the threat of the other colliery (Ashton) being closed, served to deepen the prevailing depression for Ashton people in a unique manner. It is in this creation of a common set of experiences for Ashton people that the particular features of mining in Ashton reinforce the primary influences on social solidarity. In addition, the backwardness of welfare developments in the Ashton Colliery (see above) –

6

there are no baths and only inadequate canteen facilities–is part of the reason for a general belief that Ashton is neglected and something of a backwater. Always one of the first things mentioned when the status of the miner comes under discussion among miners or among non-miners is the fact of his having to travel from work 'in his muck'. Miners are naturally rather ashamed at having to do this, and the removal of this low-status-mark from the miner by the widespread building of pithead baths has not yet reached Ashton. The existence of baths at neighbouring collieries thus again marks off the individuality of Ashton.

In the history of miners in this country a certain incident taking place in Ashton takes deserved prominence. After six weeks of the great lock-out of 1896, the striking miners of Ashton Colliery had a dispute with the manager because surface-workers were loading the inferior brand of coal for sale. On September 6th the manager, Mr. A. J. H., agreed that the loading of this 'smudge' should cease for a period. On the following day 200 miners, protesting that the agreement already had been broken, overturned a number of loaded trucks. When Mr. A. J. H. applied to Castletown for police protection, he could not be obliged because 269 police were on special duty at the St. Leger race meeting at Doncaster. A visit to the Chief Constable at Wakefield resulted finally in the sending of a telegram to York and the drafting to Ashton of a Captain Barker and twenty-eight infantrymen. All was quiet on their arrival at 4 p.m., but there soon gathered a large procession which marched to the colliery and sent a deputation to demand the withdrawal of the soldiers. By 7 p.m. the military contingent was more or less besieged in the engine-house and the crowd outside, now throwing stones, was in an ugly mood. Captain Barker agreed to leave and his unit reached the railway station, followed by the crowd, only to be turned back on the arrival there of Mr. B. H., a Castletown magistrate. By the time the troops arrived at the colliery again, a large crowd had gathered. Mr. H. proceeded to read the Riot Act of 1714, making it a felony for the crowd to remain together for more than one hour thenceforth. After this period had elasped, at about 9 p.m., the troops were ordered by the magistrate to open fire. Two men, J. A. D. and J. C., were shot dead and sixteen others injured. By 11 p.m. reinforcements arrived and the crowd dispersed.

The event is still remembered by old people in Ashton and it does, of course, fit in with their general experience of the place of the miner in our social history. However, the event has for Ashton itself no great significance. By the younger generation the event is known only vaguely and is thought of as part of the general history of 'hard times'. A 17-year-old youth–a miner's son– when interviewed hazily grouped together the years of unemployment, the shooting of the two men, and the General Strike in 'the days before the war', and it is in this general sense that the event is remembered, apart from the atmosphere of excitement recalled by those old people who were then children.[1] That the events should occur in Ashton was an accident of history. The lock-out in Ashton was part of a wider fight and similar disputes were common at the time. Questions were asked in the House about the drafting of extra police and detachments of troops to areas as far from Ashton as Somerset and north Wales. It was a chance event that Ashton should become the focus of these forces and events working out on a much wider canvas.

In fine, those features peculiar to mining and miners in Ashton itself add no new quality to the characteristics of the miner already described. But the participation in and sharing of a common set of community relations and experiences through time gives confirmation to those characteristics, considerably strengthening them.

[1] It appears to be a social fact of general significance that past experiences in a community's history take on a framework determined by the function of the memory in the present day as much as by the actual order of their happening. Social anthropologists have often noted the coalescing of time in the limited historical knowledge and mythology of the peoples they have studied. The clearest and most convincing examples of this tendency is the knowledge of genealogies possessed by societies organized on the basis of unilineal descent; see especially E. E. EVANS PRITCHARD, *The Nuer*, Oxford University Press, 1937.

CHAPTER III

Trade Unionism in Ashton

THE nature of work in Ashton explains in large measure the nature of trade unionism in Ashton.

A miner's work is quite different from that of a factory hand. Factory production depends upon uniformity of the product, for it is only then that the two great advantages of factory production can operate. These are the full development of the division of labour, and the smooth flow of the product through the various specialized processes.

The previous chapter has shown that the task the miner is called upon to do varies constantly as the conditions under which he works vary. The result is that though the factory hand from time to time may wish to secure a change in the wage rate he has contracted to receive, given the rate for the job at any particular time, there are few occasions for dispute. It is quite otherwise with the miner, whether he is a contract-worker at the coal-face,[1] or on the roads, or a day-wage worker at the coal-face or elsewhere underground.

The reason why unionism has such a powerful appeal can best be shown by recounting the kind of disputes which arise in any and every week, and in which the men concerned turn to the union for support. The most fertile source of dispute is wages, and the contract-workers are the men most often affected. The contract-worker's wages are made up in the following manner. First there is that part of the wage composed of the prices he has received for the various jobs he has done in the week, and which were provided for in the price-list. The filler working on any particular face knows that he will receive, say, 2s. 4½d. for every ton of coal he shovels on to the moving face-belt. If there is a band of stone in the coal seam then this will probably be provided for in

[1] A face-worker is defined for the purposes of official statistics as one "working on coal-face and roadways within 20 yards of the face excluding those on gate-end loader".

the price-list. Each of the various jobs which the drillers, machine-men, drawers-off, panners, rippers and the rest have been described as performing, is provided for in the price-list in a similar way. The price-list is, therefore, a formidable document often covering from 40 to 100 items. Even then, many circumstances arise which are not included in the price-list. Payment for these constitutes the second element in the contract-workers' wages, those items which have been negotiated on the spot between the workman and the management (represented by either the deputy, the overman or the undermanager). In these cases the workmen may agree, for example, to work for a certain price the coal on a face which has developed a fault. Sometimes, however, the contract-worker may not be able to earn his usual wage by fulfilling these piece-work contracts even though he has reported for work every shift. To allow for these cases a minimum wage of 24s. 11d. per shift[1] has been fixed. Thus if a contract-worker was unable (for reasons other than his own inefficiency or slackness) to earn on piece-work more than say, 10s. in a particular shift, another 14s. 11d. would be automatically added, to make his wage up to 24s. 11d. for that shift.

The minimum wage is not a matter of dispute at branch level, for it is not amenable to change at that level. It has been negotiated by the National Union of Mineworkers and National Coal Board, meeting together in a body called the Joint National Negotiating Committee for the Coal Mining Industry, and can be changed by that committee only, or in the event of the two sides failing to agree, by the National Reference Tribunal.[2]

TYPES OF DISPUTES IN MINING

Certain important disputes do however arise at branch level. Conditions are so various and unpredictable in the mine, that the job described in the price-list often corresponds in only a rough and ready way to the job actually to be done. The miner may, therefore, disagree with the management on this point, the management perhaps contending that the job done comes under a different

[1] In 1953.
[2] *Third Report of the Board of Investigation into Wages and Machinery for determining wages and conditions of employment in the Coal Mining Industry.* H.M.S.O., 1943.

category on the price-list from that which the man claims, or it may be claimed by the management that the job done is part of the job which is already shown on the price-list and for which he has been paid the agreed rate.

Examples of this are to be found frequently in those cases in which the coal-face, after being undercut by the coal-cutting machine, has been insufficiently blown to the ground by high explosives. This sometimes happens when the hole the driller has made for the charge of high explosives partly collapses before the charge is inserted. The coal is then much more difficult for the collier to shovel on to the belt, for it has not been broken by the explosives and has to be broken by the collier himself. The collier will claim an allowance in view of his difficulties, while the management will claim in most cases that such difficulties are already allowed for in the stated 'filling price' per ton of coal.

It is in those cases where the jobs are not provided for at all in the price-list, however, that the contract-worker most feels the need of union support. In this he resembles the dock-worker, who does also depend to an unusual extent on what are called 'spot settlements',[1] that is, a settlement which has been made on the spot between management and men when unexpected contingencies arise in the course of work. Payment for work not provided for on the price-list gives rise to several grievances. One of the most common of all is the allegation that the price shown on the pay-note is less than the price agreed upon. The usual explanation offered by the management is that the discrepancy arises from the failure of the worker to fulfil the contract—that he has either fallen below the level of efficiency required or failed to complete the task. Another common reason for appealing for the union's help is that the price agreed upon was too low for the workman. In these cases there is not much the union can do about it. Sometimes the deputy, overman or undermanager will not offer a price until the job is completed, the formula used being: "Do the job and I'll pay by results", an arrangement which is taken by the man concerned to imply that he will be dealt with generously. When the payment is seen to be below expectations the complaint is

[1] See *Leggett Report*, Cmd. 8236. H.M.S.O., 1946.

likely to be peculiarly bitter, because with the dissatisfaction with the payment is mixed resentment at being tricked, and anger for being fool enough to trust the management to be fair.

Among the day-wage men underground there are fewer changes of circumstances to bring the men and management into opposition and, therefore, there is less occasion for the men to need the support of the union. Nevertheless these occasions are by no means rare. Because these men work for a standard day wage, it is not often that more money will be demanded for a particular job. Occasionally this does happen when a group of men find themselves called upon to perform an unusually heavy task for which help cannot be found, or on which because of cramped space only a few men can possibly work. The union is ordinarily called upon to support day-wage men in their requests that more men be allocated to a particular job.

One proof that unionism in Ashton, as elsewhere in the coal-mining industry, derives much of its strength from the unexpected circumstances which constantly arise in the course of work is that among that group which does not experience these vagaries, the surface-workers, trade unionism has never had such a strong appeal. Trade unionism is very strong among the contract-workers, stronger indeed than among almost any other group of workers in any industry, because their daily and weekly livelihood depends upon it. It is weaker among day-wage men underground where the help of the union is less often needed. It is weakest of all among the day-wage men on the surface, for they work for fixed wages under conditions which vary very little. The labourers on the surface may, therefore, feel a vague desire to better their lot, but that is not the kind of aspiration which can be realized by the action of the branch secretary in direct consultation with the manager.

The strength of mining trade unionism rests in great part then, on the fact that certain groups of miners are constantly being thrown into a bargaining position with the management. Their own position is exceedingly weak unless it is buttressed by the background threat that, if their demands are not granted, then part or the whole of the colliery's personnel will withdraw its labour.

There is, however, another source of strength. Earlier chapters[1] have shown that Ashton is a community of its own, geographically and to some extent socially separated from other communities. Furthermore in Ashton most working-men are miners. The result is that the miner who did not join the union, unpopular because solidarity is recognized as being the basis of the union's bargaining power, could be made to feel the full weight of the community's displeasure. That this displeasure was not expressed in mild and gentle, or some might say, even civilized ways, can be seen from the following advice, given by the secretary of the Yorkshire Mineworkers' Association in the Y.M.A. *Journal*, 1923. He advises the member of the trade union to tell the non-unionist:

". . . that we want his help, his co-operation, and not his hostility, in the great work which confronts us.

Tell him that his mates look upon him with suspicion, with disgust, with contempt.

Tell him he is an Ishmael, an alien, an outsider, a parasite, a social leper, a scab.

. . . If he still remains obstinate, then by all that is just and right, and sweet and clean under heaven, tell him that he must 'get'. Must clear out of the community of clean thinking men . . . that as far as you are concerned you will shun him as you would the plague."

Although there were until recently one or two non-unionists in Ashton, cases of true blacklegs, i.e. men who worked during strikes, were extremely rare.[2] In those special çases where the coal-mine is in a town, such as the Bradford Colliery in Manchester, and the miner is therefore free of other miners in his leisure time, if he so wishes, unionism is much less strong.

LEADERSHIP IN THE NATIONAL UNION OF MINEWORKERS BRANCH

The conditions which give rise to a union of this nature also explain the character of its leadership. An almost essential

---

[1] See Chapters on 'Place and People' and 'The Miner at Work'.
[2] A story is told in Ashton of a miner who "blacklegged" during the 1926 strike. Not only his workmates and friends, but even his own sons ostracized him. When he died in 1946 none of his sons attended his funeral.

requirement of a branch union official is that he should be a contract-worker. Partly this is because the competition for contract jobs results in the most forceful personalities securing them. In the main, however, union officials are drawn from the contract-workers because the contract-workers are the men most concerned about the union, and they want one of themselves to represent them. Their superior position is acknowledged by the day-wage men who consequently, not feeling the contract-workers' urgent need for someone to protect their particular interests, tend to follow the contract-workers' lead in the choice of officials.

There are two union officials in Ashton who were not contract-workers on election. One of them was elected as the men's checkweighman. This official is present with the 'company' weighman when the tubs of coal newly sent to the surface are being weighed, and he is there to ensure that true weight of the coal is obtained and recorded, for it is upon the basis of that weight that the collier's earnings are calculated. Though miners were first given the right to appoint a checkweighman at their own expense in the Coal Mines Act of 1860, it was not until 1887 that an Act of Parliament defined the rights and duties of checkweighmen as they now are. He is appointed by a majority of the men. A committee of workmen must be formed to supervise the election or removal of the checkweighman by ballot, and to fix his remuneration. Arrangements are then made to deduct at the colliery office the due proportion of the checkweighman's remuneration from the wages of all contract-workmen. An important point is that since 1886 the workmen have been permitted to appoint whomsoever they pleased to this position, thus removing the necessity to appoint a workman at the colliery concerned. The checkweighman by virtue of his privileged position has always been prominent in the union. In a famous passage in their *History of Trade Unionism*[1] the Webbs say of the checkweighman:

'It would be interesting to trace to what extent the special characteristics of miners' organizations are due to the influence of this one legislative reform. Its recognition and promotion of collective action by the

[1] B. and S. WEBB, *History of Trade Unionism*, 1921.

men has been a direct incitement to combination, the compulsory levy . . . has practically found for each colliery a branch secretary free of expense to the Union."

In Ashton in particular the 'exception' is more apparent than real, for the checkweighman was originally a collier.

The second official who seems 'exceptional' is one of the two most powerful men in the branch, yet on election he was not only not a contract-worker, but he was actually a surface day-wage worker, and it was the vote of the day-wage men which gave him his position. Shortly after the miners were nationalized the surface day-wage workers at Ashton Colliery had a grievance, the exact nature of which it is now difficult to determine,[1] though it is said that a lengthening of their working day was threatened. It is not important. What is important is that the dispute arose near to a holiday time. In the week before a holiday, contract-miners traditionally work hard in order to earn high wages. When the surface day-wage men said they intended to cease work, therefore, the underground men, who were not involved in the grievance but who would be unable to work if there was no one on the surface, volunteered to take the place of the surface men if they were paid contract-workers' rates. There was a well-remembered meeting in the Ashton Hotel at which a young man, hitherto inconspicuous in union affairs, rose and spoke at large about the position of the surface-worker. In the next union election he received a large vote and has been since that time an increasingly prominent branch official. Here again, however, the exception is not as startling as may at first sight appear. This official was once a contract-worker and came to work on the surface owing to poor health. Soon afterwards, as an official, he returned to contract-work below ground. Finally there is reason to suppose that part of his support originally came from contract-workers, particularly in view of the fact that other contract-workers were offering to 'blackleg' that is, do the work of those surface men on strike.

In practice, the leaders of the union, the branch officials, are drawn from a group narrower even than that of the contract-workers.

---

[1] The exact cause of dispute is often forgotten, though the fact that a dispute has taken place is nearly always remembered.

We have seen that the various groups of miners work in teams of different kinds. The machine-men work in teams of 4 or 5, the drawers-off and pan-turners in pair within teams, the back-rippers and drillers singly or in pairs and so on. The colliers work in the largest teams, consisting of from 5 to 50 men. Because the team's earnings are calculated for the team as a whole, then divided among the members of the team, a collier soon acquires an interest in all the others with whom he works, for it is upon their efforts as well as his own that his earnings depend. Because the teams are large, individual colliers are more widely known by name and reputation, other things being equal, than individuals engaged in other work. In addition, a collier who disputes the adequacy of the team's earnings successfully is remembered for it by the members of the team. Apart from the two 'exceptions' considered above, there are only two members of the present union branch committee who were not fillers when elected: one was and is a pan-turner, the other was a driller and is now a back-ripper.

What qualifications or qualities are demanded of a union branch official? One of the most important things is that he should be able to 'talk'.[1] This does not imply that the Ashton miners are inarticulate. Among themselves they are lively and interesting conversationalists, although within a somewhat narrow range of subjects. What it does imply is that they are usually inarticulate except among themselves and feel this keenly. What they require of their officials is not necessarily well-thought-out policy, or even a ready flow of words. What they do demand is that when the official speaks he should speak confidently, as if he knew what he was talking about. A man who qualifies his statements is thought to be on poor ground. Under the circumstances this emphasis on confidence is probably sound, and something of an unreflective adaptation to the social environment. Experience has taught Ashton miners two things. They remember that since 1926, due to the depression in the coal trade, the Tunnbridge Colliery Co., was willing to grant very little which was not forced upon it by

[1] The best talker of course, is not necessarily the best negotiator, or the best judge of the justice of a man's claim.

the fear of the greater loss which might be incurred by a strike. Faced with the owner's implacable hostility (or what is at any rate remembered as such) fine arguments were at a discount and willingness to concede an opponent's point was fatal. The second lesson is that the management side is in any case better at arguing than are the representatives of the branch. Far better, therefore, to state what is self-evidently right and be careful not to give the equivocator a chance to turn truth into falsehood by his irrelevancies.

Another indispensable attribute of a union official is a reputation for an unswerving loyalty to the interests of his members. It is not at all necessary for him to pander to every request he receives. One of the top Ashton branch officials takes pride in the fact that if he receives an 'unreasonable' complaint (that is, a complaint which *he* judges to be unreasonable) he makes it quite clear that he is not interested in it, and that it should not have been laid before him. The loyalty required is not that of a delegate who is bound to carry out the wishes of those he represents, though there are many who think along these lines, and many more who feel that since nationalization the union has tended to become overbearing in the extent to which it sits in judgment upon the members' complaints. What is required rather is that which is perhaps best described by the word they use. . . . that the official should be 'sound'–that there should be no shadow of doubt that self or any other kind of interest may lead him to take the management's side. The most heinous crime of which a branch official can be accused is that of 'saying one thing upstairs and another in the pit-yard'.[1] This attitude gives rise to difficulties when efforts are being made by the management to secure the co-operation of the workmen, especially when this is sought through their union.

Two other factors have a certain influence in determining who is chosen for a branch position. Although drinking beer is normally regarded in Ashton as a manly pastime, and is in one of its aspects the badge of maturity, a union official who is anything but a moderate drinker stands in danger of losing his job. One former official, a good talker according to Ashton standards, and 'sound'

[1] That is, saying one thing to the manager and another to the men.

to the point of being dishonest[1] on behalf of the men, lost his position on the union committee for just this reason. What was permitted and even admired among the men at large is looked at askance in a representative of the men.

While this is true, it is also true that the same officials tend to be elected again and again. It has often been remarked upon that working-class organizations adhere with extraordinary constancy to a man who has once been elected an officer. This is due, perhaps, partly to a generous objection to 'do a man out of a job' and partly to a belief that any given piece of work can really be done as well by one man as by another. It is also conditioned by the belief that elected officials may be corrupted; thus, if an official has so far not become corrupted, why take a risk with someone new? This derivative of cynicism about elected officials is a great force of inertia.

NATIONALIZATION AND THE FUNCTION OF THE BRANCH

The effect on mining trade unionism of the nationalization of the industry in 1946 and the subsequent changes, has been complex but not profound. While the actual changes in ownership and technique have been welcomed and appreciated by the miners, the trade union itself, which played such an important part in securing the changes, has been regarded with increasing dissatisfaction.[2] This is not due to mere perversity or ingratitude. It is associated, at colliery level, with the appearance of a real divergence of opinion as to the true function of a union official.

The preceding analysis has shown that miners require the aid of their union branch primarily in order to secure from the management those conditions of work, especially the wages, which they

[1] He is the men's 'clotcher'. This is a position somewhat similar to that of a check-weighman. The tubs are sent to the surface containing stone (clod) as well as coal. A sample of the tubs is taken, and the actual proportions of stone to coal determined, payment being only for the coal, with perhaps a small allowance for some of the stone. The checking is performed by a representative of the management in the presence of a representative of the men. The men's 'clotcher' at Ashton Colliery has a reputation for being able to prove that 'a tub of stone is a tub of coal'.

[2] Among manual workers in the coal-mining industry, there is practically 100% membership of the National Union of Mineworkers. In a sense, therefore,' the union itself' refers to the total manual labour force of the mines, and no distinction can be drawn between 'the men' and 'the union'. In this context, however, what is meant is that the hostility of the 'men' is directed against those who carry on the day-to-day work of the union, namely, the officials at all levels.

feel they are entitled to expect. It has shown too that the strength of interest in the trade union branch among the different groups in the mine depends on the degree to which their demands are liable to be disputed by the management. Finally it has shown that the trade union branch is principally concerned, taking wages as the leading example, not with the absolute level of the wages in the industry as a whole, but with the narrower object of obtaining 'the rate for the job'. The rate for the job is that rate which has customarily been obtained for similar jobs in the mine concerned in the past, and is being paid at present in mines in the neighbour- hood. This being so, it is not surprising that the local branch has as many grievances to deal with nowadays from contract-workers who earn 50s. a day, as it had in the years before nationalization from contract-workers who earned only 10s. a day. It is simply that conditions of work in mining have remained the same in essentials, so that just as many disputes arise over the prices of the various jobs which go to make up the 50s. as arose over those which went to make up the 10s. Similarly there are as many dis- putes as ever in which day-wage men need the help of the branch, for they arise out of the nature of the work. In this crucial respect, therefore, nationalization has not changed the primary function of the branch, nor has it diminished the need for the branch to exer- cise it. As far as the workmen are concerned they need the union to do the same job, and to do it to the same extent; it is needed, so they think, to represent unswervingly their interests in opposition to the interests of the management.

Although the wages question is by far the most important one as far as the men are concerned in their relationship with the union, other questions do arise. In all cases what the men want is that their union representative should act as their advocate. More than that, they want him to support them even though they may know that they have broken some clearly laid down rule of long standing in the pit. They want the union to act as counsel for the defence, and to assume the attitude of 'my members, right or wrong'.

Before nationalization it was not difficult for the union to play this part. Indeed, it was rarely given the opportunity to play any other. Opposition to the management[1] was its very essence, and notoriously, the management reciprocated in kind. Industrial relations were here based on:

> ... the simple plan
> That he should take who has the power,
> And he should keep who can.

There was only one way in which the union official might fail to do as the members wanted. The official would always pass some sort of judgment on the members' grievance. He would always consider whether or not it was 'reasonable'. It was 'unreasonable' if there was no hope that it would be rectified if raised with the manager. What would be the use of arguing a hopeless case. It would lower the union official in the esteem of the manager, and this would be neither pleasant for the official personally nor useful for the official as representative of his members. The simplicity of the plan was scarcely modified by this consideration, however, for it was still the interest of the men as distinct from the interest of the management which determined the direction of the decision. It was not a question of the good of the mine as a whole. It was not a question of deciding possible effects of the decision in those spheres where the interests of management and men were identical. Such identity of interest, generally speaking, was not acknowledged to exist.

Under these circumstances, the union branch was judged not only, and sometimes not at all, by what successes it achieved in remedying grievances, for the intransigence of the management was well known. Judgment was based on the kind of effort made. Before nationalization that effort was straightforward, simple, unambiguous and therefore easily understood. It was unqualified opposition. A good example of the way in which the union in

[1] The opposition of interest as between management and men in the matter of wages is a view accepted by both sides, and allowed for in the machinery for fixing and adjusting wages. The degree to which the interests of the two sides are antagonistic in this matter, and the degree to which they need be is discussed elsewhere. The fact is that they are regarded as being antagonistic by both sides, and negotiations are conducted on that supposition.

Management is used throughout this chapter to mean all those on the management side, from deputy to owner, and not just the manager. Indeed, relations with the manager himself were often quite good.

those days could emerge from a dispute having achieved nothing, yet incur little or no criticism from the men in the process, is provided by the lock-out at Ashton Colliery in July 1938.

Of the two collieries at Ashton, Manton Colliery and Ashton Colliery, the former had been closed permanently in 1935 after a dispute over the percentage of dirt (stone) being sent to the surface among the coal. In Ashton it is firmly rooted in popular belief that the agent of the colliery caused it to be closed, saying that he would "make the women and children of Ashton bleed". In 1938 the management threatened to close Ashton Colliery also. At the beginning of July 1938 the management complained that since January the amount of dirt sent to the surface had increased from 7% to 17% and as a result of this £27,000[1] had been lost. In future, the maximum percentages allowed were to be 6% for dirt where the coal was won by hand, and 7% where the coal was machine-cut, otherwise the colliery would be closed. To this the union branch replied that the fault lay with the management. When machines had been installed, it had been agreed that when the coal had been cut, the large quantities of dust usually formed in the process of cutting would be removed before the fillers came on to their shift. This had not been done. As an alternative to the management's demand, the branch suggested maxima of 12%, 10% and 7%, according to conditions. These proposals were rejected by the management. A ballot of the members was then taken to decide whether to accept or reject the management's proposals, and the result was 750 for rejection and 259 for acceptance. The management thereupon served 7 days notice, to expire on July 16th, to all employees at the colliery. On July 15th the management repeated its terms, which were again rejected by the branch. On the next day, therefore, all workers at Ashton Colliery were dismissed their employment with the exception of a few who were retained for 'drawing off', the process of closing a colliery permanently.

Meanwhile the question had been discussed by the Urban District Council, of which the branch secretary was a member. It was resolved that the U.D.C. should ask the management to meet a deputation of the Council and of the union branch. There is no

[1] This was the figure publicly disclosed by the management.

evidence that the branch secretary played any great part in having this resolution passed. He seems to have confined himself, even there, to blaming the management for the trouble and denying that the colliers were at fault.

As a result of this resolution a meeting was held between the colliery company, the Urban District Council and the Ashton Colliery Branch of the Yorkshire Miners' Association on July 20th. The meeting continued on July 22nd and the branch officials emerged with an undertaking that the management would honour its former agreement to clear out the machine dust before the colliers came to work. On the basis of that undertaking the branch recommended a return to work on the terms which the Company had offered in the first place, namely, a maximum of 6% dirt when the coal had been won by hand and 7% for machine-cut coal. The men had risked the complete loss of their livelihood in Ashton and in the event had achieved nothing. Yet they had no doubt that the branch officials had acted as they ought; no one concerned lost his place on the union committee because of it and no one today maintains that the men's stand on this occasion on the advice of their branch officials was a foolish and wasted effort.

Nationalization destroyed the unambiguous and simple position of branch officials and substituted for it one which is far more complex. While the possibilities of achieving the branch's aims are increased during the new dispensation, so are the occasions for misunderstanding between the men and their representatives. Simple opposition is easily understood and a certain glamour adheres to it. Co-operation is always more complicated and it is not easy to reconcile the miner to the idea of co-operation with the management. This difficulty can be illustrated in its extreme by reference to the remark of a 32-year-old collier at Ashton colliery. The industry's superannuation scheme[1] was being discussed and he said: "It's a rotten scheme and you won't catch me having anything to do with it. Anything the management wants the men to do is bound to be to our detriment. That's what I've always been

---

[1] Payments are: Workmen 1s. 6d. per week (surface-workers 1s. 3d.); N.C.B. 2s. per week per member (surface-workers 1s. 8d.).

Benefits are: pensions to supplement old-age pensions ranging from 10s. to 30s. per week according to length of service and regularity of attendance. By the end of 1951 300,000 mineworkers had applied for membership.

7

taught." As has been argued above, the miner's experience in the pit confirms him in his belief that co-operation is a chimera, for is not the union called upon by him only on those occasions when the management is trying to deprive him of his just deserts? The position the branch is trying to maintain, however, is even more complicated than that. It is attempting to blend the two contrary processes of opposition and co-operation.

BRANCH LEADERSHIP, THE MEN, AND THE MANAGEMENT IN 1953

Under nationalization two sets of rules were instituted to govern the relations between management and men. One set is that concerned with joint consultation, and is designed to secure the co-operation of the men in the efficient and safe working of the colliery.[1] There is a Colliery Consultative Committee[2] consisting of the manager of the colliery, three officials he appoints to sit with him, a deputy's representative elected by the deputies, and six representatives of the men[3] elected by the men. One of the assumptions under which the Colliery Consultative Committee operates is that the men have a valuable contribution to make to the safe and efficient running of the colliery, by virtue of their practical knowledge of where inefficiency and danger in fact lie. The Coal Industry Nationalization Act 1946, states that: "The policy of the Board shall be directed to securing consistently with the proper discharge of their duties, the benefits of the practical knowledge of persons (in their employment) in the organization and conduct of the operations in which they are employed." The other assumption is that the men have an equal interest with the management in the matters with which the C.C.C. deals. Two features of the C.C.C.'s working follow from this. Items for the agenda may be submitted either by the branch or by any person

[1] Only the colliery consultative machinery is dealt with here. There is also machinery for consultation at Divisional and National level.

[2] These were first set up in 1940 ('Pit Production Committee') but they soon languished. They were revived in 1941 with scarcely more success. When the Ministry of Fuel and Power was formed in 1942 however, they were again revived and persist today largely in the form then given to them.

[3] Each group of workers elects its own representative or representatives; thus there are two representatives of the face-workers, one of the haulage-workers, one of the surface-workers and so on.

employed at the colliery, and decisions are arrived at by agreement
– there is no voting. This leaves the manager free from dictation
in carrying out his necessary duties, by allowing him, as chairman,
a wide latitude in his interpretation of what has been 'agreed'.
Nevertheless, this opportunity to play a part in the running of
the mine has not been neglected by the branch. It is something
which was long desired by the men, and which the management,
until these committees were set up, conspicuously refused to
grant. The president of the branch has been an active member
since 1947, and in 1953 attended all of the 24 meetings held. In
1953 the secretary of the branch attended 13 out of the 24 meetings
and another member of the branch committee attended 20 of the
24. It is not too much to say that the function of the C.C.C. is as
important in letting the men's representatives hear the point of
view of the management as it is in letting the management hear
the point of view of the men. The manager, with his wider pers-
pective which easily includes the view that higher production or
less absenteeism, for example, benefit both sides of the industry, is
enabled by consultation to convey this view to the other committee
members. Unfortunately, it seems to go no further than the
members. The result is that consultation, which was intended to
bring the men into closer contact with the management, has been
successful to some extent in persuading the men's representatives
(including the branch officials) of the virtues of co-operation. But
in doing so it has tended to alienate the representatives from the
men without reconciling the men to the management.

The same may be said of the set of rules concerned with the
conciliation machinery, that is, the procedure which governs the
settlement of matters regarding wages and conditions of work.
Since nationalization[1] grievances of this nature, have been dealt
with in the following manner. First of all a grievance is reported
by the man concerned to his immediate superiors, in most cases the
deputy or undermanager. If the cause of grievance is not remedied
within 3 days then the man is entitled to see the manager or his

---

[1] The present conciliation machinery was that suggested in the Third Report of the
Board of Investigation into Wages and Machinery for Determining Wages in the Coal
Mining Industry (the 'Greene Board') and adopted by agreement between the Miners'
Federation of Gt. Britain and the Mining Association in 1943.

representative about it. It is only if it is not remedied by him that
the branch is called in to aid the complainant. A Pit Meeting[1] is
arranged between representatives of the management and of the
union. It can be called for within 3 days of failure with the manager;
it must be convened within 5 days after it is called for, and its
work must be completed within 14 days. If agreement still cannot
be reached, then the grievance is transferred from the colliery for
discussion by a Disputes Committee, composed of representatives
of the North-East Divisional Coal Board and of the Yorkshire
Area of the National Union of Mineworkers. In the matter of
wages and conditions of work it is frankly conceded that there is a
clash of interest between the management and the men. In the
event of failure to agree, therefore, the problem is submitted for
arbitration to a District Umpire.

Through this machinery the branch officials' constant and sus-
tained contact with the management[2] is maintained. They are
enabled to carry out their job more efficiently. But here again,
while the opportunities for influencing the management have been
multiplied, so have the management's opportunities for influencing the branch leaders. Also, the advantage of the right of having
a grievance dealt with within a specified period implies the right to
procrastinate for the full period. The tendency therefore has been
for the management to take full advantage of the opportunities for
delay when it has suited its purpose, and for the branch to do
likewise. Because the Disputes Committee is expected to deal with
questions referred to it from any of the 115 collieries in the North-
East Division, because disputes at the collieries are frequent, and
because agreement at colliery level is often not possible, grievances
can lie for several months waiting to be dealt with by that body.

Another change which has been introduced since nationalization
though of a more 'informal' kind than those introduced in the
consultation and conciliation procedure, has had similar effects in
reconciling the union officials to the management and alienating

---

[1] "Pit Meeting" is the term used to describe this particular stage of the conciliation
procedure.
[2] The machinery, with the exception of the Disputes Committee is not very different
in practice from the procedure which operated informally before nationalization. The
difference lies in the fact that the branch now has the definite right to meet the manager
and negotiate a particular matter within a specified period.

them from their members. Both effects can indeed be seen here in an exaggerated form. It has already been shown that union officials are almost invariably colliers when elected. It has also been shown that some of the best contracts are usually those secured by back-rippers and men on roadway development work.[1]

A list of the present jobs of the ten branch committee men in Ashton speaks for itself.[2]

(1) Development.
(2) Development.
(3) Development.
(4) Back-ripper.
(5) Back-ripper.
(6) Back-ripper.
(7) Back-ripper.
(8) Back-ripper.
(9) Checkweighman.
(10) Panner.

(It is perhaps significant that the panner was newly elected at the time of this investigation.[2] Twenty-eight years old, he was elected on the demand that all union policy should be discussed in the open at the branch committee or general meetings, and not decided by groups of three or four members of the union committee who believed they were able and powerful enough to do so. "Union business belongs in the Welfare,[3] and not in the Crossways[4] or on the pit steps." Nine months afterwards,[5] still a panner, and still criticizing the deficiencies of his colleagues on the committee, he had a fight in his 'Workingmen's Club' with one of the older established branch committee men, and cut him rather badly about the face with his fists. Fighting in the clubs is usually strongly deprecated by those connected with them. The sequel of this fight, however, was that he was asked to become a member of every other club in Ashton.)

[1] Some types of development work are better paid than others, e.g., work on roads is better paid usually than on the development of new coal-faces.
[2] Branch election, June 1953.
[3] The Miners' Welfare Institute where branch general and committee meetings are held on alternate Sunday mornings at 10.30 a.m.
[4] The Crossways Hotel.   [5] March 1954.

There are various reasons why the management should facilitate and the branch officials should acquiesce in the transfer of union officials from the coal-face to contract-work elsewhere below ground. Branch officials often find it necessary to attend to union business during working hours: if they are working at the face in a team then the most important production cycle in the colliery may be disrupted by their absence; their absence from back-ripping jobs and development work is far less inconvenient. A second reason is that men working on these jobs can be paid very generously for the work they do without affecting very much the total wage-cost at the colliery, for they work in pairs or small groups, often remote from the rest of the workmen. Both the management and the branch officials accept the practice of arranging generous contracts for union officials on this work because they want to avoid any suspicion of 'victimization'. It has always been convenient for the branch officials to transfer to jobs in which their absence on union business did not hinder the efficiency of the colliery. Before nationalization, however, this usually meant taking a less remunerative job, and even, on occasions, leaving contract-work altogether and becoming a day-wage man. The well-paid contract jobs in which branch officials now soon find themselves are partly a token that the management is not at all hostile to the union and that a union official, in contrast with pre-nationalization practice, will lose nothing by taking office.

However, this generosity can also be explained by the simple argument that a representative of the employees will be more intractable if the grievances he is seeking to rectify are also his own. A dispute involving, for example, a group of colliers who claim that they have been underpaid will normally be more vigorously pursued by a union official who is himself one of the colliers concerned than by one whose own pay will not be affected by the issue. Another consideration is that an employee who is being well treated has less reason for seeing the union and the management as necessarily antagonistic forces. The fact that with only two exceptions all the branch officials are now back-rippers or on development work is almost invariably interpreted by the rank and file miners in terms of the last two arguments, implying that they have been given these jobs so that their militancy as branch officials will be impaired. Because it is so obvious that the branch officials are now in those comparatively easy, well-paid jobs, and they are thought to be in them for these two reasons only, much of the dissatisfaction with the union is crystallized around this point. It will be said of an official: "Look at X. Y. He couldn't burst his way out of a wet paper bag, yet he takes home a big pay-packet every Friday."

A well-known and malicious story runs as follows:

"The Branch Secretary was absent from his work one day. The man put in his place filled only 3 tubs during the shift yet was paid 57/- for doing so. This man is said to have 'gone about the clubs shouting his mouth off about it'.[1] No one has since been permitted to take the Secretary's job in his temporary absences."

This story may be apocryphal. It is nevertheless told and believed because what it related is possible.[2]

It is the influence of the officials above them in the union hierarchy, however, which is the most powerful factor persuading branch officials to adopt a conciliatory policy. Furthermore this influence is reinforced by the fact that it lies with the area union[3] to grant or withhold permission to strike. The branch itself has not that power. In Yorkshire this has been the case ever since 1863.[4] But whereas the Yorkshire Miners' Association was willing to permit strikes at the branches where the branch members voted in favour of such action, since it became the Yorkshire Area of the National Union of Mineworkers not one strike has been authorized.[5]

This is not because branch members no longer wish to strike in order to have their grievances remedied. The men concerned

---

[1] The Working Men's Club. There are six in Ashton.

[2] In the branch election of June 1954, the official standing for re-election was defeated by a collier. The defeated official said, "They wanted a collier, that's what did it." A miner present in the group discussing the official's defeat caustically replied that now the collier was an official, he wouldn't remain a collier for long.

[3] Since 1944 this has been the Yorkshire Area of the National Union of Mineworkers. Before that it was the Yorkshire Miners' Association.

[4] The South Yorkshire Miners' Association was formed in 1858 and the West Yorkshire Miners' Association in 1863. In 1881 they were amalgamated to form the Yorkshire Mineworkers' Association.

[5] The question of why this should be so is too large to be included in a discussion devoted to trade unionism in Ashton.

simply strike without authorization (unofficially). More than that, the men often strike for the purpose of forcing the area union to take cognizance of their grievances. That this is true of Ashton will be shown below. It is true throughout the country. A report on a dispute at the neighbouring colliery of Upton provides a typical example: "... The men held a meeting last night and decided to continue the strike until officials of the Yorkshire Area of the National Union of Mineworkers visit the colliery to take up their grievances with the management."[1]

This kind of dispute can be seen passing into open hostility at the area meeting of the union at Whitburn Colliery, Co. Durham. On February 26th, 1954, 1,800 men stopped work there because of the 'tyrannical and bullying attitude' adopted by one of the overmen towards the men in his section of the colliery. Ten days later the striking miners were joined by 2,200 more from the Harton and Westoe collieries. The strike was not ended until March 20th, after the Whitburn men had been out for 23 days, and the Harton men, as well as Westoe men, for 13. Of the strike the secretary of the Harton Branch (or 'Lodge' as it is called in Co. Durham) said: "We have not been fighting the Coal Board; we have been fighting the Union."[2] The secretary of the branch is here seen supporting the strikers and opposing the policy of the area union.[3] What happened was the men did not strike against the advice of the union but that the union branch called a strike against the advice of the area and national officials of the union.[4]

This raises an important point. It has already been said that the predicament in which the miners' union in Ashton finds itself today is not only that it has abandoned a policy of single-minded opposition for the more complicated one of co-operation, but that it is trying even more to pursue both policies together. To be more accurate the branch has no settled policy, but vacillates uncomfortably between the one extreme and the other. Partly this is because the branch officials, while being convinced to some extent of the

---

[1] *Manchester Guardian*, June 3rd, 1954.   [2] *Ibid.*, March 22nd, 1954.
[3] The Durham Area of the N.U.M. formerly the Durham Miners' Association.
[4] The strikers were called upon to return to work by Sir William Lawther, Mr. W. Jones, Mr. S. Watson, and Mr. A. Horner, but were quite unmoved by their appeal. When the Durham Area of the N.U.M. met to discuss what was to be done one of the branch officials was asked what he thought of the meeting. He replied, "I couldn't care less."

merits of co-operation, are by no means entirely convinced. One reason for this is that the belief in the feasibility of co-operation, derived from their work on the various committees set up since nationalization, has not erased the effects of their pre-nationalization experience. Nor does their present working experience convince them. The result is that the branch officials have a curiously ambivalent attitude towards these committees. Their work on them is usually characterized by a sense of responsibility for the efficiency of the colliery as a whole, even when this means treating individuals among their own members harshly. Thus at a meeting of representatives of the men and management at Ashton Colliery early in 1953 it was said: "I do not think it right that a man should be signed on again after he has been dismissed for absenteeism." This was not an undermanager speaking, harrassed by the disruption caused by absenteeism, but the branch secretary himself. On another occasion (in August 1953) the president of the Ashton Colliery branch could be seen protesting at the lack of trust the agent of the colliery had in him. A group of surface-workers stopped working one day and asked to see the president to discuss their grievances with him. The agent of the colliery went to them and told them that, "The branch president was not the manager and they could either get on with their work or go home." The president subsequently complained to the agent that when he (the president) had been sent for in the past he had used his influence in keeping the men at work, to which the agent replied that he would always be grateful for the president's assistance when it was required.

There is ever present, however, a latent hostility ready to manifest itself at the slightest sign that these committees are a mere façade. The ambivalence of the branch's position cannot be better illustrated than by contrasting the above account of the president's activities with his often expressed and no doubt genuinely held opinion that "the sole purpose of all these committees is to weaken the branch".

There is no question of dishonesty. Each of these contradictory views is held with equal conviction.

The granting to union officials of well-paid back-ripping and development jobs also has the effect of 'softening' the representatives. However, a constant limiting factor here is the long period spent by most of the branch committee members at the coal-face before their election.

The final reason for the persistence of the traditional anti-employer attitude in the statements of branch officials to the men is the most important. The men demand firm opposition to the management, and their representatives must prosecute it. To return to our last example, the president would not have been called upon by the surface-workers if they had thought that his intention would be to "use his influence in keeping the men at work". The men themselves have no firm belief that such could be his intention, though the feeling has been growing that it may be. (The president himself, however, is thought to be an extremist in the other direction. It is said that he is a Communist.)[1] On all occasions it is necessary for the officials to speak to the men in terms only of a fundamental antagonism in the relations between management and men.

The first few months of 1953 saw a rambling succession of minor disputes over loss of earnings through bad conditions, and by the summer the ground was well prepared for a serious clash. In the dispute of August 1953 which followed there emerged very clearly the ambivalence of the branch officials' position and the exasperation and confusion of the rank and file miners. This dispute, in the Beeston seam at Ashton Colliery, was caused by the installation of a conveyor-belt to replace wagons (or tubs as they are called) on the main haulage roadway. In the Beeston seam three and sometime four coal-faces were being worked in 1953. Before the new conveyor was installed each coal-face had its own system of haulage,[2] for the conveyance of coal from the face and of supplies (steel props, etc.) to the face. In the Beeston seam this system consisted of face-conveyor-belts emptying on to another conveyor-belt (loader-gate belt) leading away from the coal-face to the main haulage road. The loader-gate belts then emptied into tubs on the main haulage road, which was common to all the faces. The full

---

[1] He is not, yet he does not attempt to scotch this belief.
[2] The coal-face with the haulage system which serves it alone is called a 'district'.

tubs, drawn by a cable, worked by a large electric motor, to the bottom of the shaft, were then raised and weighed, the account of the coal from each face being kept separate from that of other faces, and wages were paid on that basis.

In July and August, during the two weeks annual holiday, the tubs on the main haulage road were replaced by a special kind of conveyor-belt called a cable-belt. This meant that it was no longer possible to calculate by weight at the surface the separate output of each face,[1] since all the coal from each of the faces was loaded underground on to the same cable-belt. It was therefore necessary to abandon the system of paying the men by the weight of the coal raised to the surface, and pay them according to the volume of the coal removed from their coal-face in the course of the shift.

This innovation gave rise to a number of grievances. In the first place, the calculations of earnings by reference to the volume of coal instead of by weight deprives the men of the protection afforded by their checkweighman.[2] Calculation by volume is felt to be in any case more complicated than calculation by weight. The men have always thought of coal in terms of tons and hundred-weights. They are now asked to think of it in terms of cubic feet. One hears constantly such phrases as: "It's difficult for a working-man to get hold of cubic measure." In fact the president of the branch once attributed the whole trouble to this one cause. "There is no one below ground who can work out all this cubic measure. That is the top and bottom of it."

Underlying this attitude is a deep-seated suspicion, amounting in some cases to a definite conviction, that if the men do not know exactly what they are entitled to, then advantage will be taken of their ignorance.

Secondly there are all the grievances attached to the proposed new price-list which the change of working hours has required. The men dispute the validity of the management's contention that 29·2 cubic feet equal one ton of coal. That this question was allowed to crop up as a grievance at branch meetings for months is sufficient warrant of confusion. There is, of course, no question

[1] In the traditional system, the total wage due to all the colliers at one face is calculated: this is then divided equally among the individual colliers.
[2] See p. 89.

of 'fairness' or 'unfairness'; it is simply a matter of fact. The specific gravity of Beeston coal is such that a ton of it either does or does not occupy 29·2 cubic feet. It is strange that the branch officials did not arrange to have this matter settled by a competent person very early in the dispute, unless they wanted some argument, however specious, to fall back on at difficult points in discussion with the management. The temporary price-list which was negotiated between the management of the colliery and the branch was not accepted by the men. The old price paid for shovelling one ton of coal on to the face-conveyor-belt was 2s. 4¾d. when the coal was machine-cut, and 3s. 10½d. when the coal was cut by hand. The new price-list was based on linear measure along the face. Machine-cut coal was to be paid at a rate of 4s. 2½d. per linear yard, and hand-got coal at the rate of 6s. 7½d. per linear yard. The other dimensions of this yard of coal were 4 feet 6 inches in breadth (i.e. the distance the coal-race advanced each day) and 3 feet 6 inches in height (i.e. the height of the coal seam – the variation in the height of the seam being, of course, allowed for in the calculations). In order to compare the new price with the old, therefore, a slightly complicated arithmetical exercise is required. The new price is in fact an improvement, the machine-cut coal, for instance, now being paid at a rate of 2s. 7d. per ton (if 29·2 cubic feet are taken to weigh one ton and when the seam is taken to be 3 feet 6 inches high) as compared with the old rate of 2s. 4¾d. per ton. Here again, however, time is constantly taken up at the branch meetings in arguments as to whether or not the price is higher. This is partly because this question is involved with another. At the coal-face some dirt (stone) is always mixed with the coal due to falls of roof, and it is very tedious for the collier to sort it out and dispose of it underground. When the tubs were sent to the surface, selected, usually early in the shift, the stone and coal separated, and the actual quantity of each weighed. The ratio for all, and the men's earnings, were calculated on this basis. The colliers accordingly judged when the 'clotchers' were at work and tried to avoid putting stone among the coal. In this way they were able to ensure that some of the stone they filled when the clotchers were not checking would be counted as coal. This mildly corrupt practice was automatically eliminated when payment by volume was substituted for payment by weight, for the volume of coal only is measured.

Mounting suspicion, lack of clarity amongst the men as to the real situation, the failure of their representatives to give that clarity, and the by no means unimportant sequence of difficult conditions in the Beeston seam, finally came to a head in the threat of a strike. There follows a report of a general meeting of the branch about two months after the cable-belt had been installed.

"About sixty men were present altogether, sufficient to crowd the normal branch meeting-room.[1] The men in attendance all had the same grievance – they were dissatisfied with the changes brought about by the introduction of the cable-belt. Most of the men concerned were present, and most of those present were those immediately concerned. Rank and file members are typically interested in their union only in so far as it represents their interests in immediate disputes with the chance of getting results.

"All those who spoke were blunt and forceful, often condemning the previous speakers' suggestions vigorously and roughly, though without rancour. Nobody waited for the chairman's permission to speak. Out of a rabble of voices one emerged and was listened to. The chairman lightly tapped his pencil on the table to call for order. The meeting was disorderly, but not because he was incapable of controlling it. It seemed that he did not wish it to be otherwise.

"Old grievances cropped up but were dismissed with irritation by others as having been dealt with a month ago. But nevertheless they obtruded. Apparently on one face the coal-cutting machine was removed and replaced by pneumatic picks on the grounds of safety. Some of the men did not like it. An official said that the men were consulted about it and had agreed to it. 'Maybe we did,' the speaker answered, 'but the compressed air pipes were being laid a fortnight before the men were asked at all.' This question being allowed to subside, the men's memories turned then on the sins of the undermanager, and someone demanded to know how he could still be in his job after his culpable neglect of the left-hand air intake of one of the faces.

"The more immediate grievances, however, took up most of the time. Because of the installation of the cable-belt the Beeston colliers were 'taking home only 25s. and 30s. a shift. That is no good at all, no man with a family can live on that'. (No one pointed out that all who

[1] Only a small proportion of the colliery's employees – an average of 30 out of some 1,500 – attend branch meetings. What happens is that a man who has a grievance will come to a meeting in order to get the branch to remedy it.

work at the mine other than contract-workers rarely receive any more than this.) The low wages of the Beeston coal was harder and lighter[1] than other coal, someone said, yet pay is not proportionately higher. The main complaint, however, was that wages since the installation of the cable-belt had been considerably lower.

"The secretary of the branch ventured to lay some of the blame for low wages on the men themselves. Before the cable-belt was installed, he says, the average per collier per shift was from 10 to 12 tons. It has dropped to 7 or 8 tons and last week it was only 5 tons."

The following exchange ensued:

"Member: 'Five tons! We've been getting out 8 and 9 tons.'
"Sec.: 'Five tons! Go upstairs[2] and look at the figures. I'm talking about last week.'
"Member: 'Well, I'm talking about the week before.'
"Sec.: 'Well, I'm talking about last week.'
"Member: 'Well, I'm talking about ...'"

It is a little unusual for an official to openly appear to side with the management by not siding with the men. It soon became clear, however, that here the motive was unexceptionable. "Yorkshire" (i.e. the Yorkshire Area of the N.U.M.) he added, "would be behind us if the men in the Beeston pit were above reproach. But five tons. . . . You would really have to get stuck in and show that a living wage could not be earned however hard you tried."

The attitude of the other officials was more clear cut. Here are a few statements made by officials at the meeting. They are given verbatim.

"We will support any recommendations the Beeston men want to make, however extreme."

"The Beeston men keep coming out, then allow themselves to be talked into going back – and achieve nothing." (i.e., they ought to be more determined.)

"What is the use of the 'go slow'? You might as well be on strike."

Quite apart from the matter of the dispute under discussion, incidental remarks were passed by the officials which underlined their belief in the fundamental hostility of management and men:

---

[1] Differences between the specific gravity of different types of coal are not great. The idea here apparently is that the lighter the coal the more shovelling is required to shift a given quantity.

[2] i.e. the manager's office. Upstairs at the colliery, not upstairs at the Miners' Welfare Institute where branch meetings are held.

"Experience shows that if the management suggests any change it works eventually to the detriment of the men."

These statements can be contrasted with the other statements opposite in tone, and can be set against the fact that the branch has not the power to support its members in "any action, however extreme" – the branch *qua* branch cannot support a strike unless it is permitted to by the area. The opposite view can also be seen clearly expressed in this account of a conversation with a union official. He said that the men's ignorance of cubic measure was the main cause of the Beeston dispute. In reply it was suggested that a more important reason was that payment by volume definitely meant that no dirt would be paid for. The official's answer was that the branch officials knew that such was the case, but the rank and file did not: "We dare not tell the rank and file. We have enough trouble keeping them at work as it is."

Thus alternately agitated and restrained, uncertain of whether they were being supported or opposed by their leaders, the Beeston colliers struck work, and returned to work, engaged in 'go-slow' tactics, and then tried to earn as much as they could under the existing price-list. In early September 1953 the colliers on three faces in the Beeston seam withdrew their labour and were joined by colliers from another seam, striking in sympathy – 3,200 tons of coal were lost. In late September there was another strike. In mid-October they adopted a 'go-slow' policy and caused a loss of 4,300 tons. Near the end of October 1,235 tons were lost because of the colliers on two faces 'going slow' and 2,040 tons because of a strike. In November a sit-down strike cost 850 tons, and 730 tons were lost during a 'go slow'. After that, with Christmas approaching and because many of the most dissatisfied colliers had taken jobs at other collieries, the Beeston seam became quieter. In January 1954 there was another strike involving 69 colliers, and 1,200 tons were lost. By now, to the grievances connected with the installation of the new cable-belt, was added another. The Beeston colliers were complaining that 'dead work' there was not being paid for. By 'dead work' they meant those jobs in the mine which arise outside the normal cycle of production, for instance an unexpected fall of roof may block one of the roads and this road block will have to be removed with pick and shovel. The men in the other seams feared that alleged attempts to cut allowances for 'dead work' if successful in the Beeston seam would be extended to the other seams. Furthermore it was said that: "We've all got to come out on strike before Barnsley[1] will bother to do anything about it for us."

The result was that in February 1954 workmen in all the seams of Ashton colliery ceased work. This had the desired effect. The secretary[2] of the union for the Yorkshire Area came to Ashton. What did he say? "What could he say?" a disappointed collier remarked after the meeting. "He said we should go back to work to allow the conciliation machinery to operate."

CONCLUSIONS AND GENERAL CONSIDERATIONS

Serious studies of 'industrial relations' in this country consist for the most part of detailed description of the development and functioning of the formal negotiating and conciliation machinery in the industry concerned. The foregoing account suggests that a great deal of light may be shed on such problems by taking into consideration the less formal relations at the level of the individual enterprises. At the level of day-to-day industrial relations in Ashton, for instance, there is no doubt that the theoretical method of procedure and settlement between management and men is very far from meeting the demands of the situation. In particular, those persons in positions of responsibility for the prosecution of the 'new' harmonious type of industrial relations, the trade union branch officials, are by no means adapted to their 'new function' (we do not suggest by any means that representatives of management are perfectly adapted to the promotion of equal and harmonious methods of settling disputes, but our descriptions and analyses are here concerned only with trade unionism). For example, the fact that branch officials now drift into well-paid jobs, when in the past they were open to victimization, is not capable of explanation only in terms of local industrial relations. It appears to be a phenomenon by no means confined to the area being

[1] The Yorkshire Area of the National Union of Mineworkers Headquarters is situated in Barnsley.
[2] At that time this was Mr. W. E. Jones.

described and certainly it is reasonable to suppose that such features reflect general policy in a centralized, in this case a nationalized, industry.

It is suggested that the present patent contradictions in trade unionism in Ashton are the result of a development set afoot certainly no later than 1939. This development manifests itself, as we have seen, in the widening split of interests and policy between union officials and rank and file. Its secondary consequences within the industry are serious; there are innumerable 'unofficial' strikes, largely inconsequential, and therefore bound to lead to greater dissatisfaction both with the industry as such and with the union leadership. In a particular community such as Ashton the frustrating and confusing character of these tendencies is evidenced time and again in the recent history of worker-management relations. To summarize the attitude of Ashton miners, and that of many others; the union is a prime necessity, and yet it tends more and more to refuse to represent their view of the essence of relations with management. In fact it is common for union leaders to condemn strikes in the same terms as the management.[1]

In some measure the history and structure of the union itself, and that union's relations to British society as a whole, explains the arrival of the present impasse. More than sixty years ago in Yorkshire a number of local and county organizations of mining trade unionists had grown up, specifically as militant bodies with officials whose duty was to represent the interests of their members against the employer. Over the past sixty years in particular these local and county organizations have come together and have developed the formal characteristics of a large-scale bureaucratic organization.[2] The paid officials of such an organization will not inevitably develop interests divergent from those of their members, but their position in such large-scale organizations certainly exposes them to influences far removed from the experience of the ordinary miner. Not only is the full-time official relieved of the

---

[1] Although only a half-dozen men in Ashton itself are winding-enginemen, the best remembered example of this feature is the 'Winders' Strike of December 1952, when the N.U.M. sought to supply substitute winding men from areas outside Yorkshire to maintain production.

[2] This development is characteristic of many other trade unions, cf. V. L. ALLEN, *Power in the Trade Unions*, London, 1954.

daily physical toil and danger of the miner's task, but, the higher his position in the union the more removed he is from the whole of the routine life of his previous workmates, of his members. The potential for such a divergence in the case of a miner's union is especially high. In the first place, mining villages and their conditions are remote even from the ordinary industrial town, and bear no comparison with the pattern of life to which a full-time official, say in London, must adapt himself. Secondly, mining has been visited more than any other industry with the attentions of the official administration of the country's economy and government. Since 1939, with the first strict war-time control and the Essential Works Order, and then nationalization, with the mines brought still closer to the organs of public power and administration, the level of coal production, the maintenance of the mining labour force, and most important from our present view-point, the preservation of 'industrial peace', have been direct concerns of the Government. There has resulted a proliferation of committees of all kinds, and a whole series of new duties of trade union officials, at all levels. These, together with the growing possibility of union officials being appointed to posts in the management of the industry give rise to a group of officials in the union whose actual lives become more and more isolated from those of their members.

Within the union certain officials feel this growing isolation and speak half-nostalgically of the days in their native villages and valleys, but this does not solve the problem. These officials exist on salaries and with expense accounts which must be comparable with those of people with whom they have to deal from day to day; they grow used, of necessity, to the same kind of life and entertainment as other executives in bureaucratic organizations. If the determining influences in the miner's attitude to his union were derived from his daily working experience, we can only expect, given the rather moribund democracy of the National Union of Mineworkers branches and areas, a loosening of the ties between officials and men, not only through lack of acquaintance but through a different life-experience, producing a different emphasis of interests. The miner, we saw, has little prospect of 'social mobility' except through the collective effort of his mates, and this has very limited aims in the main. For the full-time official, however, the prospects are not a little different. Not only is there the possibility of promotion in the union itself, with at each level the various conferences and meetings in very pleasant places and good hotels, the chance, for those of such inclination, of coming into the public eye through public meetings, the Press and even the radio and television. In addition, men with trade union administrative experience are more and more thought suitable for posts in management, particularly in the nationalized coal-mines. Here are real prospects of individual success.

Now it is not suggested here that these tendencies are inevitable nor bound to work out in a certain way. They are simply a back-ground or framework in which policies are decided and carried out, in which individuals operate. But it is maintained that out of the experience and organization of coalminers at the point of pro-duction, in their everyday relations, there grew up a formal organization with a complex history. This formal structure, itself originally the consciously-formed weapon of one side in an 'in-formal' set of relations, has developed along lines determined not by the necessity of the 'informal' basis so much as by the 'normal' mode of development and formation of modern large-scale bureaucratic organizations. As such an organization it interacts with other like organizations—in particular the National Coal Board at all levels—and its representatives with the repre-sentatives of those organizations. In fact the personnel of the two sides becomes over a period similar to a greater degree than there is similarity between the interests of the officials of the union and its basic rank and file.

In such a situation a real alienation between the interests of union officials and union membership is latent, and in the experi-ence of many coalminers a reality. Certainly in Ashton it has become manifest. The mode of development of real everyday in-dustrial relations is found to conflict with the leaders' actions and words, which are determined by a different set of relations, those summed up in the mode of development of the union as such, in its relations with like bodies. In such bureaucratic organizations there is an inevitable inertia in policy since 'getting on' within the structure depends more and more on conduct acceptable to those executives who lay down policy.

The material of this chapter is mainly relevant to one particular moment in this contradiction. Precisely at the meeting-point of these two sets of relations, i.e. between the everyday work of the miners and the formal structure of the union, stands the branch committee. The day-to-day experience of the men at work demands the leadership of the men on this committee, yet the continual pressure from full-time officials seeks to make them men of a different mould, in this particular era to make them men with ideas opposed to those of their members.[1] It is by reason of their 'sandwiched' position in the structure of social relations in mining and mining trade unionism that the branch committee, as a group and as individuals, behave and speak in flatly contradictory ways. Their conduct can be explained only in terms of 'these overall structures of which they are part. In the structure of informal relations between man and management they are charged with the duty of militant leadership. At the bottom rung of the leadership of the National Union of Mineworkers, they are charged with the opposite duty. As things stand at the moment they cannot occupy both positions at once and behave consistently.

[1] This is not to say that trade union leaders no longer desire or fight for better conditions for their members – this they must do to maintain their membership – but rather that most leaders, at least in the N.U.M., conceive of the possibility of such improvements on a basis completely different from that conceived by their members.

CHAPTER IV
Leisure

I T might be supposed that the miner's trade union which is so important in relation to the miner at work is also important to the miner apart from his work. In fact some trade unions are important in both these ways. One of the directions in which a union may affect the lives of its members apart from their working lives is by acting as a friendly society, that is, a non-profit-making society to which members make regular weekly subscriptions. In return members or their families receive financial help from the society in the event of such exigencies as sickness, unemployment, old age, and death.[1] There are two ways in which this extends the influence of the union over the leisure time of its members. On the one hand it helps bind him to the union while he is at work in a manner which is divorced from his desire that it should help him in disputes with the management. It makes him think of his union as affecting his whole life and not merely his work. Indeed the concern of the union and its members with 'friendly' benefits may militate against the efficient fulfilment of its function with regard to the members' conditions of work. One of the most succinct statements of this potential cleavage was that contained in a resolution of the Congress of General Railway Workers in 1890. "That this union shall remain a fighting one, and shall not be encumbered with any sick or accident funds."

In Ashton the union branch does not extend its influence into the leisure time of its members in this way. Such 'friendly' benefits as are paid to members are administered not by the branch itself but by the Yorkshire Area of the National Union of Mineworkers. These benefits, as compared with some unions, are

---

[1] Serious setbacks have been suffered by friendly societies since the creation by the Government in 1946 of its own machinery to administer an 'all in' scheme of National Insurance. Membership of friendly societies fell by nearly 2,000,000, i.e. 21 % from 1947 to 1952. Total funds, however, rose from £203,730,000 in 1947 to £215,996,000 in 1952. (*Report of Chief Registrar of Friendly Societies for 1925*, H.M.S.O., 1954.)

not large.[1] The expenditure per member of the Yorkshire Area of the National Union of Mineworkers is smaller than that of many unions, especially those described as 'craft unions'–those admitting as members only workmen who have acquired the appropriate craft skill, usually after a long apprenticeship. This expenditure in the first half of 1953 consisted of a weekly average of £1,864 on friendly benefits, spread over a total membership of 144,557.[2] Of this sum a weekly average of £1,496, equivalent to 3s. 4d. a week for each pensioner-member was spent in providing pensions. The remainder, £368 per week, was spent on funeral benefits.

For some unions, particularly in the past, the weekly payment of the friendly society dues to the branch or lodge secretary became a social occasion for members.[3] In Ashton not even the payment of the union dues can now provide an occasion for weekly meetings with the branch secretary and at least some members. Nowadays union dues are deducted by the management before the miner receives his pay.

There is one final reason why the Ashton branch must be regarded as too closely concerned with the miner at work to be regarded as an important part of the framework of leisure. For the miner in Ashton, as elsewhere, joining the trade union is all part of starting work. The management, by arrangement with the National Union of Mineworkers, asks 'new starters' to undertake to join the union at the same time as they agree to start work at the

---

[1] In 1948, for instance, the Amalgamated Society of Woodworkers spent 49% of its total income on friendly benefits, while the Amalgamated Engineering Union spent 44% of its total income for that purpose. (*Annual Report of the Chief Registrar of Friendly Societies*, H.M.S.O., 1950.) The Yorkshire Area of the National Union of Mineworkers in the first half of 1953 spent only 29% of its total income on friendly benefits.

[2] Membership of the Yorkshire Area of the National Union of Mineworkers as at June 30th, 1953, was:

Full Industrial Members	124,547
Half Industrial Members	6,490
	131,037
Pensioners	8,987
Widows	4,510
Members of H.M. Services	23
Total	144,557

[3] These payments frequently took place in public houses. Many friendly societies are still closely connected with the public house, e.g. the Royal and Antediluvian Order of Buffaloes.

colliery. The union is therefore from the first primarily identified with their jobs in the men's minds.

LEISURE INSTITUTIONS AT THE COLLIERY

The colliery itself is only of slight significance in leisure activities. One of the most important organizations stemming from the colliery is the brass band, the Ashton Workmen's Band. The word 'workmen' is not intended to indicate that it is workmen who play in it (though in the main that is the case) but that it is the workmen who finance and manage it. There is a levy of one penny per week on the wage of all workmen at Ashton Colliery – that is one penny from each man's wage is deducted by the management on behalf of the band and is credited to the band funds. In the case of these 'levies' it is theoretically possible for a workman to apply at the colliery office for exemption from this deduction – to 'contract out' – but it is rarely done in practice.[1] It is called a 'workmen's band' to differentiate it from what are called 'company' bands, i.e. bands which are sponsored by the management.

The difficulties with which this band has had to contend are in many ways typical of the difficulties which face all those associations which depend for their success on the continuous efforts of their members. The fact that different members of the band work on different shifts – 'days', 'afternoons', or 'nights' – and that each individual member may change his shift from time to time, means that the regular meeting together of all members is impossible. In the case of the colliery bands the position is somewhat easier than for most associations. Where the band is sponsored by the 'company' it is usual for all bandsmen to be given work on the day shift. The case is then very similar to that of the union officials[2] – the bandsmen are put into jobs away from the main teams of men, and paid at rates which compare well with what they would be able to earn on any other shift. The Ashton band is not a 'company' band

---

[1] A miner's pay-note, apart from the usual deductions for National Insurance and income tax also carries a section entitled 'Variable Deductions'. These are 'voluntary levies' in favour of which the union branch has voted. Once the vote is taken, dissenters rarely apply for exemption. A large number of them have no idea what some of the levies are for. At present the 'Variable Deductions' on an Ashton Colliery pay-note are numbered 1–13 and it is necessary to apply at the colliery office in order to discover to what the numbers refer.

[2] See Chapter III, 'Trade Unionism in Ashton'.

however and the shift problem is ever present. The manager of the colliery will usually grant any request which the bandmaster may make for a bandsman to change from one shift to another. This may happen, for instance, in the week before a brass band competition, when the band want to practise every day. But the management cannot always guarantee that the bandsman will be placed in as remunerative a job as that which he had on his original shift, though such an arrangement is attempted. The workmen, therefore, often look askance at a temporary change of shift for the convenience of the band. Another consideration which weighs with the workman is that he has to leave his team-mates for a time and work with comparative strangers.

Where the colliery band is considered important as a 'morale-booster' the bandsmen are permanently on shifts convenient for band practice – as in the case of company bands. In Ashton the workmen's band is not considered important enough to be granted this special treatment.[1] The additional advantage to be derived from putting all bandsmen on 'days' (i.e. the shift lasting from 6 a.m. to 1.30 p.m.) is felt by the management to be out-weighed by the disadvantage of arousing the jealousy of the other workmen. Despite the lack of preferential treatment the band manages to struggle along quite successfully.

The colliery band enters fully into the life of the town. It provides entertainment for the people of Ashton. The band will play at charity concerts, at garden fêtes and carnivals, at parades. People feel it is 'our band'. Its appearance is expected at minor celebrations organized in, and affecting only quite a small neighbourhood of two or three streets. There is, however, another function which it is called upon to perform, and that is to represent Ashton to the outside world. In this it resembles some other associations, the most important of these being the football team. Ashton's prestige is involved in the numerous competitions in which brass bands customarily engage. For most people the satisfaction of hearing of the success of the band is the limit of their

---

[1] There are twenty-five members of the band apart from the bandmaster. Of these, eighteen are Ashton Colliery employees, three are miners from neighbouring collieries, one is a railway worker, one owns his own business (he is a cobbler), and two are schoolboys. At one time in 1953 the soprano cornetist was the daughter of a collier.

interest. Ashton people know nothing of the competitions in which 'their' band does badly. But if it does well there is wide-spread comment.[1]

One other leisure activity closely connected with the colliery deserves particular mention–the Ashton Division of the St. John's Ambulance Brigade. It is viable as an association because it has a clearly defined and socially approved purpose directly relevant to the needs of the town, namely, the administering of first-aid to the injured. Moreover, because of the dangerous nature of mining its members can feel that what they are learning today may well be required in practice tomorrow. The local Division of the Brigade therefore has an advantage over many other associations (for instance an Evening Institute or the Workers' Educational Association) whose purpose is to train their members in some way. The member of the St. John's Ambulance Brigade can never feel that his training is pointless and remote from reality. In addition to the Division itself, which meets in a building of its own, there are ambulance classes held on the colliery premises.

Apart from their primary function both the Brigade and the Ambulance Class engage in more purely recreational pursuits. Thus the Division might organize a public dance at the town baths. This not only helps to raise funds for the organization, it also gives members and their friends an opportunity to mix freely. The resulting social contacts, rendered easier by previous acquaintance at first-aid meetings, have a beneficial effect upon social relationships in subsequent meetings. The colliery ambulance class similarly holds a social evening for its members. Prizes will be presented to members who have reached certain levels of proficiency, and the manager of the colliery may be invited to attend in order to present them. Speeches will be made, stressing the importance of the work the members do.[2]

There is also an Ashton Colliery cricket team, and there have

[1] In 1953 the band achieved a success in an area contest which led to an invitation that the band should be one of those to represent Yorkshire in Class 'B' of an inter-county contest. Other successes included 1st prize in the annual contest of a County Band Society, and 1st prize and gold cup at an inter-county contest.

[2] At a typical meeting of this sort the manager urged all young men to join the movement; one of the undermanagers gave illustrations of the value of first-aid knowledge; the branch delegate praised those connected with the movement; and two other colliery notables said that they "would be pleased to help in any way".

been from time to time (though there are none at present) colliery football teams. The difficulties presented by shift-working in cases like this are, however, almost insuperable, and teams are organized only to collapse after a session or two.

STATUTORY WELFARE PROVISIONS

Coalmining is one of the few industries in which the provision of leisure facilities by employers is required by law. The Mining Industry Act of 1920 established a Miners' Welfare Fund. A Miners' Welfare Committee was constituted at the same time to administer the fund ". . . for purposes connected with the social well-being, recreation and conditions of living of workers in and about coalmines".[1] The income of the fund was to be provided from a levy of a penny per ton of saleable output–the so-called 'miners' magic penny'.[2]

In 1926 a levy on mining royalties of one shilling in the pound was imposed for the purpose of constructing pithead baths. This money was also administered by the Miners' Welfare Commission, as the Miners' Welfare Committee came to be called. The total receipts from both these sources between 1920 and 1944 amounted to £24 million, equivalent to 22s. per annum per person employed in the industry. Of this, £6·7 million[3] was spent on the provision of 400 pithead baths. Ashton Colliery was not one of the lucky ones. The explanation given by the Ashton miners is that before 1935, when there were two collieries in the town (Ashton Colliery and Manton Colliery) the Miners' Welfare Commission offered to provide pithead baths at one of the collieries only, and the Ashton miners were to agree between themselves which colliery it was to be. No agreement was ever reached, so the baths were never installed.

In carrying out its statutory duty to provide for the 'social well-being and recreation' of the miners, the Miners' Welfare Commission spent £5·95 million between 1920 and 1944 in

[1] Mining Industry Act, 1920, Section 20 (1).
[2] In 1934 this was reduced to a ½d. because of the poor conditions of the industry.
[3] This figure includes not only the income from royalties levy but from other sources, e.g. income from the royalties levy was augmented by appropriation from the output levy for this purpose.

schemes connected with Miners' Welfare Institutes.[1] One of these
schemes was the Miners' Welfare at Ashton.

The 'Welfare': general functions

It is possible that the original failure of the colliery company to
provide leisure facilities for its workpeople has been made good
from funds which the colliery has been compelled by law to contribute for that purpose. There is no doubt that some facilities have
been provided–the only trouble is that the miners do not use
them to any great extent. The Miners' Welfare Institute is by far
the biggest building in Ashton. It was built in 1897 to accom-
modate a Working Men's Club and a theatre.[2] The building was
purchased in 1925 by the Miners' Welfare Commission, and its
'running costs' were then to be provided for by means of a weekly
deduction from the wage of each miner at Ashton Colliery.

The institute serves Ashton as a minor centre of leisure activities
in the following ways:

1. The local Dramatic and Musical Society uses it for a fortnight in each year.
2. Other organizations will put on occasional concerts.
3. Each Saturday throughout the year a dance is held on the
   premises.
4. It is used by the Boys' Club and Boxing Club.
5. It is used for billiards.

Its large hall and stage provide facilities for the local Dramatic
and Musical Society which produces one or two shows a year.
Although each show runs only for a week they are staged in
elaborate fashion, great effort being expended to secure the appropriate costumes, scenery, and orchestra. The Society's productions
appear to be largely confined to what we may call 'classical'
musical comedy, e.g. The Country Girl.[3] These shows are seen by

[1] It is customary for the capital costs to be borne by the Welfare Organizations while the
costs of running the institute are borne by the miners in the form of weekly deductions
from their pay.
[2] On the ground floor there is a billiards room, four other rooms and an office backstage.
There is also a room which is used by the Boxing Club.
[3] "Although a small mining town organization the Ashton Dramatic and Musical
Society lacks neither talent nor versatility, and is able to move from musical comedy to
pantomime with facility." (Local newspaper comment.)

approximately a thousand people, that is, an average of 200 at each performance.[1]

The Miners' Welfare Institute hall and stage are also used by other organizations when wishing to raise money or obtain publicity.

A good example of the former is the concert which was organized in March 1953 in order to raise funds for the Ashton Hospital. There is in the town an "Ashton Hospital Comforts Fund Committee". Its object is to "provide patients with the comforts the State cannot provide". In 1952 it raised £107 from donations, and £129 from whist drives, a cricket match, a concert in the Miners' Welfare Institute, a collection in a pit tub on 'Carnival Day',[2] and by various other 'special efforts'. The money was spent mainly in providing the hospital patients with a television set, a radio set and speaker, and other Christmas presents. The main work of the committee is done by the colliery workers who are members of it – the union branch delegate is the chairman, and the secretary is a clerk, though some of the doctors in the town are associated with it.

The Comforts Fund Committee Concert was held one Sunday evening between 6.30 p.m. and 8.30 p.m. and was attended by approximately 250 people. It resulted in £28 10s. 3d. being added to the funds. The Ashton Workmen's Band played *Ashton* (a march composed by a local musician), excerpts from Offenbach's *Orpheus in the Underworld* and other pieces. The concert was mainly, however, in the old music hall style of a series of 'turns', the artists being those who were under contract to appear for the evening at the various Working Men's Clubs in Ashton, and which the Working Men's Clubs 'loaned' to the Comforts Committee free of charge for the concert. After the concert at the Miners' Welfare Institute the artists then returned to perform at the Working Men's Clubs which had originally engaged them.[3]

The 'turns' were for the most part somewhat unprofessional.

---

[1] cf. Success of the Arts Council theatrical touring companies in Welsh mining areas.

[2] i.e. the annual carnival organized by the Ashton Urban District Council when people were invited to throw coins into a 'tub' (a wagon used for coal haulage at the colliery). £17 2s. 8d. was contributed.

[3] The Working Men's Club as a leisure institution with its custom of holding concerts on Saturday evening and Sunday midday and evening, will be dealt with below.

One strictly amateur item was the contribution of the Ashton Labour Party Ladies' Choir. This consisted of Victorian and Edwardian ballads. The general impression of the concert was that of an inferior music hall show.

An example of the Miners' Welfare Institute being used for publicity is provided by the concert held by the Ashton Urban District Road Safety Committee in April 1953. About 300 people were present–a large proportion of them being school children, and they were 'treated to a largely musical programme'. There were then two competitions, one was a 'Road Safety Quiz' in which school children from town and neighbouring mining towns competed for a Road Safety Challenge Cup. The other was a game of 'Twenty Questions'–this well-known parlour game being used to propagate knowledge of the Highway Code.

The institute hall is also used occasionally by local associations which wish to entertain either their own members or some other group. For example in December 1953 nearly 200 old miners (pensioners) were entertained to tea at the institute by the Ashton Colliery Branch of the National Union of Mineworkers.

Apart from being the scene of the occasional concerts and festivities of associations which are ordinarily centred elsewhere, the institute is used by some organizations as their regular meeting-place. The union branch committee meets there each fortnight on Sunday morning, and there is a branch general meeting on the intervening Sunday. It is the centre, too, of the Ashton local Labour Party. The local Labour Party itself rarely meets more than four or six times a year, but the Ladies' Section meets fortnightly: this association will be examined more fully below.

The 'Welfare': dancing

Each Saturday evening throughout the year the Miners' Welfare Institute is used for ballroom dancing. (At each dance there are 400 or 500 people between the ages of 15 and 22, the great majority of them between the ages of 17 and 20.) At Easter, Whitsuntide, August Bank Holiday, and Christmas, more than one dance is held. The number of dances held there at Easter 1953 is typical of such times. There was dancing from 7.30 p.m. until

11.30 p.m. on Saturday night. There was another dance early on Easter Monday morning from 12.15 a.m. until 4 a.m., and a third dance at 8.15 p.m. on Easter Monday which lasted until 1 a.m. on Tuesday morning. The popularity of this leisure activity among young people in Ashton can be gauged from the fact that in 1951 the total Ashton population in the age group 15–22 was 716.

The dances are organized by an ex-overman from Ashton Colliery.[1] His advertisements are phrased in the following terms in the local Press: "Come and have a Riot of Fun with George and his Boys", "Saturday Night is Riot Night". When attendance is invited in these terms it is not surprising to find that the local Court of Summary Jurisdiction records show many cases of obscene language, assault, disturbing the peace, and so on which occur at these Saturday night dances. The number of persons actually involved in such disturbances is a small minority of those actually present. However, the fact that disturbances of this nature do occur is indicative of the level of normal behaviour on these occasions.

The main functions of these weekly dances appears to be that of bringing young people together in a manner which facilitates the approach of the two sexes. Judged in this light–and not, as is so often done, in the light of the 'Riot of Fun' which it seems to the outsider to be–it is seen to be not inappropriate. In the first place the custom of the dance hall permits a young man to ask any young woman to dance, without previous acquaintance of any sort, though asking a woman who is already obviously partnered is frowned upon. The young man can approach quite without embarrassment because a first request is scarcely ever refused, except when the requests come from someone conspicuously abnormal by local standards. Also, the strain of 'making conversation' until a basis of common interest is discovered is reduced to the minimum. The common interest is provided by the dancing itself and involves not much more than both the partners being familiar with a few stereotyped dance steps. The fact that the dance floor is nearly always so crowded after an hour or so that dancers

---

[1] He is the dance band leader–that is, he is the dance band manager attending to all its affairs. The band has been highly successful, being placed fourth in an all-British dance band contest in 1953.

can do little but shuffle around means that even the basic requirement of being able to join in the dance is not very rigorous.
Conversation under these circumstances is not urgently necessary,
and can be, and often is, limited to a narrow range of remarks on
the size of the crowd and the quality of the band. In the second
place the breaking of ties between individuals of opposite sex is
facilitated. If the partners do not suit one another then the man
simply does not ask for another dance, or the woman declines the
invitation. The importance of this lies in the greater freedom of
movement it affords. Generally speaking, strangers will enter more
readily into relations with each other if they feel that they are not
committed to anything by doing so. "I spent the happiest two
years of my life at Ashton Miners' Welfare," said a young lady,
who had grown out of the Miners' Welfare age group, but being
unmarried, still went to dances elsewhere. "You could let yourself
go there."

The 'Welfare' : minor functions

There are other leisure activities which are associated with the
institute. There is a boys' club conducted under the joint auspices
of the Ashton Colliery Miners' Welfare Committee and the West
Riding County Council. It is intended to cater for the 15–18 age
group and meets twice weekly. In practice its activities are restricted to table tennis, which is played by the six or eight youths who
attend on club nights, and to cricket and football at other times.
The sports constitute the club's main appeal, and though there are
sixty members of the club[1] only the thirteen or so who play for
the football team[2] and the eleven or so[3] who play for the cricket
team are interested members.

There is a boxing club also catering in practice for youths under
the age of 18. None of its members is a resident of Ashton.

There is a reading-room in which there are a few books, and
which is never used for reading, but in which meetings are held
occasionally.

[1] The total population of youths between the ages of fifteen and eighteen inclusive was
389 in 1951.
[2] Rugby League Football.
[3] Many of those who play in the football team also play in the cricket team.

There is finally a billiard-room containing three billiard tables. Billiards and snooker constitute the only activity for which the institute is utilized continuously throughout the week. The billiard-room is open from Monday to Saturday between 10 a.m. and 9.30 p.m. In the 1930's this was a popular rendezvous for unemployed miners.[1] The return of full employment led to a steep decline in the popularity of the institute billiard-room as a centre of leisure activities. It is used nowadays almost exclusively by youths under the age of 18 – on reaching that age they are eager to become a member of one of the Working Men's Clubs which have billiards and snooker teams. The charge is 6d. per half-hour for billiards and 8d. per half-hour for snooker. In 1951 the total receipts of the institute from this source totalled £69 19s. 2d. – which on the most generous estimate indicates an average of only 54 players per week.

In general, it can be said that the Miners' Welfare Institute does not play a very important part in the leisure time of the miner above the age of 20. The Dramatic and Musical Society uses it for a fortnight in the year, other organizations each use it for one or two nights only in the year, and some organizations use it for their meeting-place. It is used to some extent by a few youths in the Boys' Club and by youths who play billiards and snooker. Only as a dance hall is it used regularly by large numbers. It is possible to argue that a large active membership is not really important in this matter, that the reason the institute is supported by levies from both sides of the industry is in order to provide a 'cultural' centre with money which would otherwise not be forthcoming. The 'cultural' value of the activities which do take place there may be judged from the account which has been given of them. Clearly whatever the possibilities of such an artificially fostered institution may be, at present it provides little which is not provided elsewhere, and it is chiefly elsewhere that the miner in Ashton spends his leisure time.

---

[1] Even as late as July 1937, when industrial recovery was well advanced in some parts of the country, the number of unemployed in the Ashton Labour Exchange Area was no less that 1,756 or 48% of a total insured population of 3,667. Of this 1,756, 1,360 were miners. By July 1939 the percentage of the unemployed had fallen to 13·5 and of a total of 524 unemployed only 168 were miners.

The aims of the Miners' Welfare Commission have been pursued, however, by its successors. The Miners' Welfare Commission was confirmed in existence by the Coal Industry Nationalization Act 1946, which also gave welfare duties to the National Coal Board. In 1948 the Miners' Welfare Commission and the National Coal Board came together in a National Miners' Welfare Committee in order to co-ordinate their welfare functions. By 1951 it was decided to separate the functions again–without losing sight of the need for co-ordination–by forming a Coal Industry Social Welfare Organization to deal with 'social welfare',[1] leaving the National Coal Board to deal with 'colliery welfare'. The investments available to the Coal Industry Social Welfare Organization total £1,250,000 of which the North-Eastern Division of the National Coal Board has been allocated £250,000. Between the foundation of C.I.S.W.O. in July 1952 and December 1952 the organization of the North-Eastern Division spent £54,000.

The Ashton miners have not benefited greatly from these schemes. Locally the institute building itself is described as a 'white elephant': it is difficult to disagree with this verdict. Nevertheless the Divisional and Colliery Welfare Committee decided that they were justified in spending more money in improving its fabric and extending its amenities. Accordingly in 1953 the Ashton Colliery Miners' Welfare Committee asked the Ashton Branch of the National Union of Mineworkers for a greater contribution from the miners towards the running costs of the institute. After a series of meetings spread over several weeks the Ashton miners finally consented through their union branch to increase the weekly contribution of each man from $1\frac{3}{4}d.$ to $3d.$ On that basis the Divisional Welfare Committee agreed to supply

[1] The Coal Industry Social Welfare Organization is concerned with more aspects of leisure than merely the fostering of Miners' Welfare Institutes. Its field of activity covers (1) Outdoor schemes: to provide facilities for healthy recreation, with special but not exclusive reference to the needs of youngsters. (2) Indoor schemes: there are 700 miners' welfare halls. (3) Holidays: C.I.S.W.O. maintains Miners' Holiday Centres at Skegness and Rhyl. (4) Health: twenty convalescent homes are provided for out of the Miners' Welfare Fund. (5) Education: Investments of £185,000 allow ten scholarships and ten exhibitions per year from the income for "Workers in or about coalmines and their sons and daughters." The Miners' Welfare Institute in Ashton is, however, by far the most important part of its work as far as Ashton itself is concerned. C.I.S.W.O. was founded jointly by the N.C.B. and the N.U.M.

£4,500 immediately in order to renovate the institute. In the circumstances it seems likely that the resources would have been better employed elsewhere.

It is clear that, directly, neither the workplace nor the welfare organization connected with the industry play a very important part in the leisure time of the Ashton miner. Indirectly, however, the influence of the workplace is profound.

LEISURE AND INSECURITY

The pursuit of leisure in Ashton has two principal characteristics. It is vigorous and it is predominantly frivolous. Without wishing to enter into the question of which are 'better' or 'worse' ways of spending leisure time, it should be explained that the word frivolous is used in the sense of "giving no thought for the morrow". It is used in this way as a contrast to those forms of recreation which pursue a definite aim such as intellectual improvement by means of study in adult classes or discussion groups, or spiritual improvement through membership of a church.

Before considering in detail the content of the individual leisure activities, it is of some importance to appreciate why they should be of this generally frivolous nature. Briefly the hypothesis is this. The way in which the Ashton miner uses his leisure has been determined by a particular set of factors. These factors influence all miners but there are many other factors which are peculiar to particular groups of individuals, and finally, of course, there are variations of individual personality. In fact the consequences vary from behaviour based on extreme prudence to behaviour based on its opposite. What can be said, however, is that in the main the latter type of behaviour does predominate, and determines the texture of community life in Ashton. The former behaviour is found only among a minority.

Among the many factors at work a small number appear to exert the predominant influence. There is first insecurity arising from the danger of being killed at work–a danger which is more real to a miner than to a worker in almost any other industry. Secondly there is the insecurity of income based upon the miner's greater liability to injury as compared with other workers. Thirdly there is the fact that in the past the miner has lacked security of employment. Fourthly there is the insecurity based on the fact that the miner cannot be sure that he will spend his life in a well-paid job (even if he escapes injury): generally speaking, in his youth he works as a day-wage man, and returns to day-wage work in the fifties. In addition to these considerations all of which are concerned with insecurity in one form or another, there is the fact that the people of Ashton not only work together but also live in the same neighbourhood. In this Ashton differs from large towns, where people working at the same place may live in widely separate places. Finally there is the influence of shift-work on social life.

How real is the danger of being killed? At its most spectacular this insecurity bears itself in upon the whole mining community of the country in the form of some major disaster at one colliery or another every few years. Some of these disasters are of such magnitude that they are remembered by the public at large. The worst of recent years have been those at Gresford, Creswell, Knockshinnoch, and Easington. Death at the pit[1] is not, however, restricted to these great calamities, as the table on page 132 shows.

There are two extreme attitudes of mind which these facts tend to engender among coalminers. They are each clearly expressed in miners' ballads. The first shows the way in which the miner might be inclined to live only for the day. In 1882 an explosion at Trimdon Grange Colliery caused the loss of 72 lives. The conclusion to be drawn was:

"Oh let's not think of tomorrow, lest we disappointed be,
Our joys may turn to sorrow as we may daily see,

[1] "You've heard of the Gresford disaster," the words of a miners' ballad run:

The terrible price that was paid.
Two hundred and forty-two colliers were lost,
And three of the rescue brigade.

Down there in the dark they are lying,
They died for nine shillings a day,
They have worked out their shift and now they must lie,
In the darkness until Judgment Day.

Farewell, our dear wives and our children.
Farewell, our dear comrades as well.
Don't send your sons down the dark dreary pit,
They'll be damned like the sinners in hell.

Today we may be strong and healthy but soon there comes a change,
As we may see from the explosion that has been at Trimdon Grange."[1]

TABLE VI

Number of men killed in British
coalmines
1946–53[2]

Year	No. of men killed
1946	543
1947	618
1948	468
1949	460
1950	493
1951	487
1952	420
1953	401

The second ballad draws the opposite conclusion from an
explosion at Seaham Colliery in 1890:

"Death to me short notice gave,
And quickly took me to the grave.
Then haste to Christ make no delay,
For no one knows his dying day."

It is with the first attitude we are concerned here. The second
will be discussed when the place of the churches in the life of the
Ashton miner is considered. Those miners who typically excuse
their liberal spending with a remark such as "well, we might not
be here another day" may be rationalizing their behaviour, nevertheless the simple fact is that this danger of death is comparatively
great for the miner.

There is in addition to the continual danger of violent death a
constant risk of injury. The more serious type of injury may disable
the miner from working for life. In 1953, of all claims (783,000)
under the National Insurance (Industrial Injuries) Act, 1946, well

[1] A. A. LLOYD, Come all Ye Bold Miners. London, 1952.
[2] National Coal Board Annual Report and Statement of Accounts, for years 1946–53,
H.M.S.O.

over one-third were in respect of persons employed in the coal-mining industry.[1]

The existence of these claims would seem to indicate the necessity for thrift on the part of the miner. What the miner feels, however, is that all the savings he could possibly muster would make little financial difference if he was seriously incapacitated and rendered permanently incapable of work. Spread over a period of several years the spending of the savings would scarcely help. Apart from saving with this prudential object, all other types of saving are discouraged. "We have tried to save before, time and time again," the 46-year-old wife of a filler said, "but something always happens and it all goes. Last time he (the husband) broke his thigh, and 'compo'[2] isn't enough to keep a family on." Her point of view was that should the family have some money 'put by' then it would inevitably be used to fill the gap between the husband's normal wage and the benefits he receives under the National Insurance Acts. The fact that his normal wage is now high makes this all the more likely, for generally speaking, people do not take easily to a sharp decline in their standard of living.[3] If there were no savings, however, "they would struggle through somehow, they had always done so before". The effort of saving under these circumstances often seems wasted.

The miner's position in this respect can be contrasted with that of other wage-earners and salaried workers. The miner is far more liable to see his savings used for purposes he never intended because he is far more liable to be deprived of his normal earnings through injury. One figure has already been quoted to demonstrate this—one-third of all Industrial Injuries payments are to coalminers. Just as telling a figure is that which shows the contrast in the injury rates as between underground- and surface-workers at Ashton Colliery. In a typical month, November 1953, the following rates

[1] First Annual Report of the Ministry of Pensions and National Insurance, 1953. OMD, 9159. H.M.S.O., 1954.

[2] i.e. Industrial Injury Benefit under the National Insurance (Industrial Injuries) Act, 1946.

[3] According to the Quarterly Statistical Statement of the N.C.B. (N.C.B. Report 1953, September) in the second quarter of 1953 the average earnings per week per wage earner in the NE. Division of the N.C.B. was 216s. and 8d. Industrial Injuries benefit is 46s. per week for a single man with additional benefits for men who are married, for children, and for dependent adults.

for accidents involving absences of more than three days were recorded. Among underground-workers the rate was 177·6 per 100,000 manshifts worked. Among surface-workers it was only 43·8 per 100,000 manshifts worked. In other words at Ashton Colliery an underground-worker, of whom there were 1,262 in November 1953, was in that month four times more liable than a surface-worker–of whom there were 473 (i.e. only one-quarter of the number of underground workers)–to be unable to come to work for at least three days because of injury.

The position of the miner can be further contrasted with that of many salaried workers in that the latter receive their normal salary during several weeks of illness. The miner is, therefore, in the unfortunate position of often losing his savings due to injury– what he calls a spell of bad luck. The tendency, therefore, is to give up saving as a bad job, and live from day to day, spending the money as it is earned in the belief that 'they'll manage somehow' come what may.[1]

Some saving does, however, take place. But it is short-term saving, and it is for a sunny rather than a rainy day. It is significant that an effort is being made to secure the adoption of a measure which would help to counteract the influence which has been discussed here. At the 1954 Annual Conference of the National Union of Mineworkers, of the four main demands put forward one was that the miner should be paid his full wages during the first six weeks' absence due to sickness.

Nor is saving encouraged by another type of insecurity– unemployment and low wages–which the miner has experienced in the past and which has had exactly the same effects as the in- security connected with the loss of income resulting from physical injury. Nowadays this insecurity is not present as a day-to-day factor. It has been at times in the past. In 1953 there was virtually no unemployment among Ashton miners and all three seams at the

[1] In April 1953 an assistant commissioner of the National Savings Committee addressed Ashton Colliery Consultative Committee meeting. He pointed with pride to the fact that in Ashton in the previous six months thirty-seven new groups had been formed and £7,500 collected. However, he added, the question was, "how were they to avoid the men taking the money out as soon as they put it in?" The president of the union branch thought "this saving would hinder production, as the men would draw their money out at the end of two months and stay off work until they had spent it".

colliery were working–that is, providing employment–on 256½ days in the year. Ashton unemployment figures for the early 1930's are not available, though as has been pointed out, as late as 1937, 48% of the insured population of Ashton was unemployed. Even those in employment, however, were unemployed for part of the year. In 1932, for instance, one of the seams at Ashton Colliery was working for only 120 days, another for 166 days, and the third for 169 days. In 1933 one of the seams was not worked at all, and the others worked only 196 days each. The hand-to-mouth existence which these circumstances often induced was described in the following terms by a 41-year-old underground day-wage worker. He was arguing that the changes for the better in the last few years were most clearly to be seen in the changed position of women.

"I remember when I was a lad, just about leaving school, I used to bring home beer from the pub in any quart bottles that could be found in the house.[1] Then my father would ask his pals in. Oh, they had right good times, you know! The next week he might have only one (shift) in.[2] He'd put 5s. 6d. on the table. It was my mother who had to manage somehow. It was her worry. My father didn't give a . . . . That's the way it was in those days! Now she knows how much money is coming in,[3] and she has the right to know. If there isn't enough, she wants to know the reason why."

The irregularity and smallness of the income of Ashton miners can clearly be seen if the earnings of three typical employed colliers at the local colliery in a rather bad year, 1931–2, are considered.

The figures show the total earnings of each collier per week at the colliery; alternate weeks only are shown (see table on page 136).

In just the same way as the dangerous nature of the miners' job has produced the two extreme attitudes of "oh, let's not think of tomorrow, lest we disappointed be" on the one hand, and ". . . haste to Christ make no delay, for no one knows his dying day" on the other, so did the insecurity of employment produce

[1] This is known as 'tramming beer'. If the father had had a 'good week in' at the colliery the boys of the family were often sent to the public house throughout the day keeping father supplied with beer.
[2] i.e. he might be employed for only one shift in the week.
[3] i.e. knows that a regular income is assured. The phrase does not mean that she knows now how much her husband earns.

TABLE VII

Weekly earnings of three colliers at Ashton Colliery in year
April 1931–March 1932

Week	Earnings of Collier 1			Earnings of Collier 2			Earnings of Collier 3		
	£	s.	d.	£	s.	d.	£	s.	d.
1	1	6	8	1	2	0	Nothing		
3	3	13	4	3	5	10	Nothing		
5	3	5	10	2	10	0	2	10	4
7	3	12	6	2	6	8	3	5	0
9	2	16	8	2	10	4	2	14	0
11	1	18	9		15	4	2	2	0
13	1	0	9		11	7	1	4	0
15	3	2	6	2	5	0	2	6	8
17	1	19	3	Nothing			1	14	6
19	3	12	6	Nothing			1	12	3
21	2	14	6	Nothing			2	4	4
23	3	3	9		19	10	3	10	10
25	2	1	6	Nothing			1	1	10
27	3	10	0	1	9	6	2	3	0
29	4	0	0	2	17	1	1	3	0
31	1	19	4	2	8	9	2	13	0
33	2	16	3	1	9	9		10	7
35	2	14	8	2	5	4	1	0	0
37	2	18	4	2	8	9	2	19	6
39	2	10	10	1	9	9	1	7	6
41	3	4	2	3	10	6	1	18	9
43	2	18	4	2	15	0	2	6	0
45	3	9	9	1	9	0	3	7	1
47	3	7	1	3	2	11	4	2	6
49	3	7	6	2	13	9	4	5	0
51	2	7	0	1	1	6	2	0	9

two extreme attitudes. One was that of serious concern over the miners' predicament and a determination to seek out its causes and cure. That attitude found expression in the work in Ashton of such organizations as the Adult School Union, the Council of Social Services, the Labour Party, and in the political and social character which the work of some of the churches assumed.[1] These will be considered later. It is with the opposite and predominant attitude

[1] Excepting, of course, the work of the Miners' Trade Union, which is dealt with elsewhere.

that we are concerned at this point; the attitude that life must be lived from day to day and whatever surplus income there is over everyday needs should be spent in securing whatever pleasures are possible. In Ashton these pleasures are mainly drinking and gambling. With the former the miner was able to escape temporarily from the consciousness of the limitations of his way of life. With the latter, if he won a moderate sum he could spend it on drinking to escape from his limitations in fantasy, and if he won a large sum he could escape in fact. Thus in 1938, when the average earnings of all coalminers were 56s. for a week of five shifts,[1] the largest Ashton Working Men's Club, with over 1,000 members,[2] had an income of £10,500 mainly from the sale of beer. The other Working Men's Clubs in Ashton, the membership of which totalled nearly 3,000,[3] had incomes which bore a similar relationship to their size.

Though the danger of injury and death in the mine is much reduced nowadays as compared with times past, the insecurity based on danger still operates. For compared with other industries, mining is still a dangerous occupation as the figures show. "They can do what they like down the pit," the miner says, "but they'll never make it as safe as working in the sun." It is fashionable to argue that since unemployment and low wages are things of the past for miners today, then surely to cling to the old habits is irrational and a survival from the pre-1939 years. Certainly many of the miner's present habits took shape in conditions no longer prevalent; but to end the discussion at that point would be a distortion of the truth.

In point of fact, low wages and insecurity of employment *in relation to a well-paid job* form an integral part of the miner's life. A great deal has been said already about the gap between the income of the contract-workers and that of the day-wage men. In the second quarter of 1953, for example, the average earnings per manshift worker (in the North-Eastern Division of the National Coal Board of which Ashton is a part) was 57s. 2d. for contract-workers. For *all* workers underground it was 47s. 8d.; this average

---

[1] G. D. H. COLE and R. POSTGATE, *The Common People*, 4th ed., London, 1949, p. 643.
[2] The membership rose from 985 in 1935 to 1,219 in 1940.
[3] The membership of all other Working Men's Clubs in Ashton rose from 2,357 in 1935 to 3,253 in 1940.

was arrived at, of course, by including the earnings of the highly paid contract men and the earnings of the day-wage men and underground-workers which are correspondingly lower. The average earnings per manshift worked for surface-workers was 31s. 5d.; again, this average includes the earnings of the surface-tradesmen, and the wage of the ordinary 'day-wage man' is correspondingly lower. Broadly it can be said that in Ashton while the earnings of contract-workers are between £14 and £15 per week, the weekly earnings[1] of the average day-wage men underground are between £7 10s. 0d. and £8 10s. 0d. and the weekly earnings of the day-wage men on the surface between £6 7s. 6d. and £7 5s. 0d.

A question often asked is why those men particularly who can apparently earn £15 per week regularly have not changed their way of life more? The answer is that the miner, generally speaking, earns low wages on day-wage work ('low' that is, as compared with what he will earn later as a contract-worker) when a young man. It is in this period that the pattern of his domestic life takes its form so that that pattern is a function of the day-wage men's average wages standard. The range of what he regards as 'necessities' under these circumstances corresponds with the traditional standard of living in Ashton. Thus there were—on a weekly average—274 workers below the age of 26 at Ashton Colliery in 1953. Of these only 43 (15·7%) worked at the face. It is not until the age group 25-50 is reached that a large proportion of highly paid workers is to be found. At Ashton Colliery in 1953 there was a weekly average of 1022 in this age group: 503 or 49% were face-workers. Having had his standard of living fixed in the low-wage days of his youth the highly paid contract-worker therefore regards much of his wages as 'free income' in the sense that nothing has a very firm claim on it. He therefore feels free to spend it on the traditional pleasures of Ashton, in the clubs, in the pubs, and in the bookie's office. And he feels free to refrain from earning it at all—in other words, to absent himself from work.

A possible objection is that the member of a profession is in

[1] In 1953.

exactly the same position in his early twenties as the miner so far as earning an income is concerned; his income is low in comparison with that which he can expect to earn later. There are several factors which help to explain the difference, some of which have been examined already. There is first the traditional standard of living in Ashton; it is low because Ashton was once poor. Secondly there is the ever-present danger that personal injury might rob the miner of his earning power; to become too used to a high standard of living which he felt was absolutely necessary might make adjustment exceedingly painful. This point does not contradict that which was made earlier, to the effect that the temporarily injured miner would spend his savings rather than suffer a cut in the amount he is used to spending on 'necessities' and 'luxuries' alike. There is a *temporary* sense of loss attendant upon the deprival of even the most trivial possession which is avoided if possible. The miner avoids this temporary sense of loss by using his 'sunny day' savings for a 'rainy day' (and on that account, as has been argued, is discouraged from saving). But by keeping his standard of living low in relation to necessities, the miner avoids the possibility of that permanent sense of loss suffered by decayed gentlemen or gentlewomen. Thirdly the miner, unlike the professional man, is liable to be temporarily 'demoted' to day-wage work as a form of punishment. There is a certain amount of temporary interchange of jobs between men on contract-work and men on day-wage work. This also tends to teach the miner to keep the standard of what he regards as necessities low. Fourthly, while the professional man expects to increase his income throughout his working life, and even the ordinary wage-earner expects to at least keep his income stable, the tendency is for the miner to be returned to the ranks of the day-wage men in the last decade or more of his working life. At Ashton Colliery there was a weekly average of 502 workers in 1953 over the age of 50. Of these only 93, or 18.5 %, were face-workers.

There are differing reasons why the miner who earns £15 a week does not change his way of life. Firstly his early conditioning does not prepare him for such a change. The weight of tradition is strong. Largely this is because even in the age group in which the proportion of well-paid contract-workers is highest–49% in the age group 26–50–there is still 51% of day-wage men. The well-paid contract-worker cannot break away from the traditions of this group because he is one of them himself, in the sense that he emerged from this group as a young man and will return to it as an old man. Secondly, because of the insecurity of his income due to risk of injury or demotion even when he is on contract-work, it would be exceedingly dangerous for him to become used to a standard of living he may not be able to sustain. The result is–and this may be regarded as a cultural adjustment to the circumstances of living–that most miners spend a large propor-tion of their income on things which are not very much missed if their income depreciates. R. H. Tawney has said that mankind hates its own prosperity and ". . . menaced with an accession of riches . . . it makes haste to pour away the perilous stuff. . . ."[1] It is not prosperity that the miner hates, but the pain of being deprived of it, and for him at any rate the accession of riches may indeed be perilous, for he knows he cannot be sure of their permanence.

Insecurity in all the forms described is the most important single factor which has moulded and still moulds the miner's way of life in those hours when he is not at work. This insecurity is not simply a sum of discrete factors, all of them adding to the total. The miner is exposed to these particular insecurities–unemployment or a rapid fall of income–by reason of his very position in the social structure of Britain. This position is in essence that of the weekly wage-earner. Because he is a weekly wage-earner the chance physical events of injury and ill-health strike him very hard, and for the miner this is emphasized by the physical nature of his job. Because he is a wage-earner he has never been able to place himself in a position to withstand those events in the economic system at large which visit misfortune upon him in the shape of depression and mass unemployment.

Another characteristic of mining as an occupation affects the form taken by leisure pursuits in Ashton. Unlike the two characteristics of insecurity and common residence (which influence in a positive way the quality and extent of community life) this sets limits to the

[1] R. H. TAWNEY, *Religion and the Rise of Capitalism*, Pelican Books, p. 85.

kind of activities which can flourish in Ashton. This influence is
the practice of shift-work, already mentioned in connexion with
the colliery band and football teams. There are three shifts, the
day shift, from 6 a.m. to 1.30 p.m., the afternoon shift from 2 p.m.
to 9.30 p.m., and the night shift from 10 p.m. to 5.30 a.m.[1]

Approximately 1,700 men work at Ashton Colliery; about 900
work on the day shift, 500 on the afternoon shift, and 300 on the
night shift. This means that the potential membership for any
leisure activity which plans to meet at some set time once a week
is only half of what it would be if all the men finished work at the
same time. The difficulties of the small group organized for some
purpose which required the regular attendance of all its members
do not, however, end there. To be obliged to draw only on one
half of the potential membership is seen to be but a slight drawback
when compared with the complications of shift-work in practice.
In the first place many of the men do not work on one shift only.
Many work on one shift all one week and on another the next,
alternating between, say, the day and afternoon shifts. Others
work all three shifts in the course of three weeks,[2] one week on
each shift. Even the comparatively regularity of alternating week
by week between one shift and another, or between each of three
shifts, however, is often broken by a workman being transferred
from one job to another, where the shift cycle is different.

PRINCIPAL LEISURE INSTITUTIONS

The trade union, the colliery and the Miners' Welfare Institute
have already been discussed and have been shown to appeal to
only a few Ashton people as centres of leisure activity. This is not
because the miner likes to spend his leisure time on his own, and
refrains for that reason from joining these organized activities. On
the contrary, the possible explanation is to be found in the fact that
the Ashton miner has developed organized leisure institutions
which adequately meet his requirements quite apart from the
facilities these other organizations provide.

[1] A minority of workers on special tasks, and occasionally the colliers on one or two faces,
work shifts different from these, e.g. beginning at 8 a.m. or 10 a.m., etc.
[2] At some collieries four shifts are worked, and in the course of a month some miners
may work one week on each shift.

The Working Men's Club

The leisure institution which appeals to more Ashton miners than any other is the Working Men's Club. There are six of these in Ashton, which, together with two small similar 'social clubs', have a total membership of 6,844.

The Working Men's Clubs are predominantly male institutions. Only one of those in Ashton admits women as members. The others absolutely forbid by rule the admittance of women into the club except for the concerts on Saturday evening and Sunday mid-day and evening. Membership is open to all males aged 18 and over who have been proposed and seconded by any two members, and accepted by the committee as being suitable. As there are only 4,824 males aged 18 and over in Ashton,[1] and the total member-ship of the clubs is 6,844, it can be seen that their appeal is wide-spread. The club membership is half as large again as the total male adult population. (This is possible because many men are members of more than one club.) The membership is not restricted to 'working men'. Of those who work at the colliery the deputies as well as the workmen are club members. This is not so in all colliery towns in Yorkshire. At some there is a separate 'Officials' Club' and deputies are discouraged from joining any Working Men's Clubs. In Ashton even overmen are members of these clubs, and one of the three undermanagers is a member of two of the clubs, both of which he regularly visits.[2]

Many of the men who do not work at the colliery are members of these clubs. Here again in addition to the ordinary workmen, many of the tradesmen are active members. Nevertheless, the miners are in a great majority. The prosperity of the clubs fluctuates with the miners' prosperity. Thus in the 1930's when there was unemployment and low wages, membership stood at less than half its present level, largely due to the fact that membership of more than one club became a rarity. The increase in miner's earnings in the last few years has led to a great increase in the membership of

[1] In 1951.
[2] "... When I'm out for a drink," he said, "a chap is liable to get a bit drunk and come over to me and start being nasty. If I'm in a pub there isn't much I can do about it. In the club all I have to do is to call a committee man, tell him what is going on, and I can be sure of redress."

the clubs as the table below indicates. New members have joined, and existing members have joined more clubs.

TABLE VIII

Club	Membership of registered clubs in Ashton 1930–53					
	1930	1935	1940	1945	1950	1953
A	1,202	985	1,219	1,320	1,500	1,550
B	640	450	883	1,100	1,300	1,370
C	572	656	884	1,127	1,248	1,232
D	70	429	350	1,046	900	920
E	250	200	224	477	855	829
F	137	245	244	483	500	512
G	—	153[1]	264	245	329	292
H	—	—	—	—	—	130
Totals	2,871	3,118	4,068	5,778	6,632	6,835

The objects of all the Ashton clubs are set down in the following terms: "The Club is established for the purpose of providing for working men the means of social intercourse, mutual helpfulness, mental and moral improvement, and rational recreation."[2]

The reality is somewhat different. The means of social intercourse are certainly provided, and there is a certain amount of mutual helpfulness, but the clubs can scarcely be said to be seriously concerned with either 'mental and moral improvement' or 'rational recreation'.

At all the clubs the bar is ". . . the centre and support . . . the pole of the tent".[3] It provides first the financial basis. The balance sheet of a typical Ashton club (with a membership of 920) for the financial year 1951–2 shows a total income of £12,500. Of this, £12,200 was 'bar takings', under £40 came from members' annual subscriptions, and the balance consisted of cash in hand.

[1] Registered 1934.
[2] Each club is required by the Friendly Societies Act to submit a Book of Rules to the Registry of Societies. Only if this Book of Rules is accepted by that office is the club able to exist.
[3] CHARLES BOOTH, Ed. *Life and Labour of the People of London*, 1902, Vol. I, Part I.

Whatever else members may do at the club, they spend a good deal of their time simply conversing over their beer. Conversation is notably free and easy. The men conversing have often been life-long acquaintances; having been at the same school and played together as children, they now, as adults, work at the same place and spend their leisure together in such places as the clubs. The clubs do not provide much variety from one to another in respect of the topics which are discussed. Within each club, however, there is considerable differentiation between the groups. A description of one of the clubs in this respect can be taken as typical.

The great majority of men who frequent this club spend most of their time at the bar, drinking and talking. The topic which surpasses all others in frequency is work – the difficulties which have been encountered in the course of the day's shift, the way in which a particular task was accomplished, and so on. A whole series of jokes are based on this fact. It is said that more coal is 'filled off' in the clubs than is ever filled off down below and that the men come back exhausted from a hard shift at the club.[1] Among some of the retired miners preoccupation with this subject is extreme. Their rendezvous in one of the rooms of the club is nicknamed 'The Pick 'oil' (i.e. 'The Pick Hole') – (the room where the picks are sharpened, stored, or used). The only other subject which is regularly discussed is sport, especially Rugby League Football. In conversation about work and football general considerations or abstractions scarcely ever appear. The discussion is almost always about concrete cases, whether of actual incidents at the colliery, or actual incidents on the field of play.[2]

There is a scarcity of conversation on the level of general principles. Any attempt to do so is dismissed as 'talk' – that is, empty argument. Credit goes to the men who 'know a lot' of concrete details.

[1] Strangers are warned to watch their heads as they enter such and such a club because of the pit props which have been erected all round, and to look out for any runaway tubs which might come careering along the rails which are to be found laid all over the club premises.

[2] Men will often go into great detail about such things as who was the deputy on No. 21 district when so and so broke his leg and at such and such a time; or who it was who scored the try that won the match so many seasons ago. Spirited discussions will often take place on what the facts actually are or were.

At most of the clubs, however, there is a 'Best Room' in which the conversation ranges much more widely, and where those who "like to talk and debate", as one of them put it, can do so. Any evening a group of between ten and twenty will be found in the 'Best Room' forming a single group, discussing the topics of the day. The following subjects were touched upon in an evening between 9.30 and 10.30 p.m. This description gives a good idea of the range of subject-matter. There were twelve men present, whose ages ranged from the 25–30 age group to the late sixties. Seven were contract men at the colliery. The other five were: a former contract-worker who was now working on the surface due to injury and old age, the storekeeper at the colliery, a tradesman at the colliery, the colliery crane driver, and the manager of the local cinema.

A recent Rugby League match between Ashton and a team from a large Yorkshire city was first discussed. This was followed by a heated argument concerning an international boxing match. The conversation at this point turned towards Government policy in regard to roads, and eventually led to the topic of housing. All these matters were dealt with from the Ashton point of view, that is, from the point of view of their relationship and importance to Ashton.

The 'Best Rooms' in the clubs and their equivalent in the public houses cater for an important part of what can be regarded as Ashton's 'intelligentsia' (the other part of Ashton's 'intelligentsia' view the clubs and public houses with disdain). These are the people who have become used to discussing matters in general terms, who generally look more often beyond Ashton and the colliery, and whose opinions are well weighed by their townsmen when they speak either in private or at meetings.[1] The Urban District councillors are almost all 'Best Room' men. The union leaders (branch committee members) use the 'Best Room' at the Crossways Hotel.[2]

It is interesting to note, however, that, even in the 'Best Rooms' where the people meet who consider themselves 'better' than their

[1] When they speak at meetings they avoid exposing themselves to the criticism of being mere 'talkers' by talking of concrete cases rather than general principles.
[2] See Chapter III, 'Trade Unionism in Ashton'.

workmates, there is a strong tendency to suppress by ridicule any attempt at differentiation.[1]

Apart from drinking and talking the most popular activity at the clubs is the concert. At the three largest clubs free concerts are held each Saturday and Sunday evening and Sunday midday. For these professional artists are engaged. The performances generally consist of a succession of songs, often taken from popular musical comedies of many years ago. The spectators sit at small tables drinking beer during the performance, and there is a 'Master of Ceremonies' to call for order.[2] Women are admitted to the concerts and do come in considerable numbers. The exclusion of women from the clubs except on these occasions results in the women having an ambivalent attitude towards the clubs. In a conversation at one of the clubs a miner laughingly said:

"If all the women's wishes for the club had been granted, it would have been blown into the middle of the Sahara desert long before now. Except on Saturday and Sunday! Oh, the club's all right then!"

The following is an account of a typical Saturday evening concert:

"The room in which the concert is held is well lit and the lights are not dimmed for the performance. The room is large enough to contain small tables and seats for about 150 people. At one end of the room there is a stage with a Tyrolean background. At the other end of the room there is a small bar.

"When the concert begins at 7.50 there are between sixty and seventy people in the concert-room. At three of the tables there are only men, but at all the other tables there are also, one, two, or three women. The men are all drinking beer, while most of the women are drinking bottled brown ale. An electric bell is rung, and the Master of Ceremonies cries 'Ladies and Gentlemen, with your kind permission D. H. will now entertain you!' D. H. sings *If I had a heart of a clown*, staring at the ceiling, beating time with his hand, and standing back on one heel. He is applauded, the electric bell is rung to signify the end of the turn and people continue the conversation which had been broken off. By 8.15 the concert-room is full. There are about 160 people present (in this one

---

[1] This is seen in even the slightest things. Someone produced a rather expensive brand of tobacco. The cry immediately went up, "My, aren't we posh", and the middle-aged collier concerned put the tin away in confusion. Again someone using a slightly unusual word such as 'proximity' will find himself the butt of 'good-natured' banter.

[2] Reminiscent of the old time music hall.

room–there are many others in the bar, in the 'Best Room' and in the 'T.V. Room'). There are as many women as men, and all age groups above 18 are well represented.

"The bell rings again, the Master of Ceremonies calls through the microphone, 'Thank you! D. H. will now entertain you. Thank you.' He rings the bell and thanks the audience several times until there is silence. D. H. sings *Memories* and the audience loudly claps his indifferent performance, stamping and shouting. The bell rings, 'Thank you', 'Some order please!' D. H. then sings *Some Enchanted Evening*. The bell rings again conversation is resumed, and there is now much going to and from the bar for drinks.

"When the singer is not performing a pianist provides background music to the conversation at the tables. It is something to listen to when there is nothing to say.

"The development of conversational groups shows an interesting pattern in the course of the evening. When the performance first began there were groups of four or six sitting at each table. Many of the groups sat for considerable periods in complete silence. What conversation there was was between couples. As soon as it was announced that the artist was about to sing there was quietness. By 9 o'clock there were very few silent groups, and all the people at each table leaned forward appreciatively listening to whoever was speaking. The Master of Ceremonies found it increasingly difficult to secure the required degree of silence. It is as if the performance was quite subsidiary to social intercourse, and was indeed used merely to facilitate it by filling in the gaps in the conversation. Later on there is constant liaison between the members of different groups. The whole audience, which was originally composed of individuals and pairs rather loosely associated in groups of four or six has become a series of groups–now somewhat larger–about six or eight–associated with other groups–in an audience now much more like a unity.

"Between songs there is a hubbub of conversation. Each group has an eruption of jollity, declining again into ordinary conversation as the impulse exhausts itself. There is a constant coming and going at the bar. Many join in the choruses as the artist sings.

"By 10 o'clock many have begun to leave the concert-room for home, and within a few minutes it is only half-full. At 10.20 it is only one-third full, but those remaining are very rowdy, and it is necessary to shout to make oneself heard to one's immediate neighbour above the din. The 'turns' have finished so there is no reason to keep quiet, and nearly everyone is more than a little drunk. There is more noise than there has been all evening. Shortly before everyone leaves, the local Salvation Army Captain, a young bandsman, and the two young girl members of the Corps, come into the room to sell copies of the *War*

Cry and the *Young Soldier*. They are received with the customary respectful indifference."

Those clubs which do not have concerts have in their place what is called a 'housey-housey session'. This is a lottery in which each player is given a card bearing a selection of fifteen numbers from 1 to 90. A 'caller' then picks tokens marked from 1 to 90 out of a hat at random, calls them out, and the first player to have all the numbers of his particular selection drawn is the winner. At these clubs women are admitted for the housey-housey sessions on Saturday and Sunday evenings, and Sunday midday in just the same way as they are admitted at the others for the concerts. Six or eight 'games' are 'played' in the course of the session, and there are usually about 70 or 80 players in each game. Each player pays 6*d.* for his or her selection of numbers and the winner wins the sixpences of all the other competitors, apart from a certain proportion which is deducted by the club for some charitable purpose – for instance to provide an annual outing for the old age pensioner members of the club. The prize for each game is usually somewhat less than £2. Sometimes the 'card' (i.e. the selection of fifteen numbers) will cost 1*s.* and the prize is then correspondingly larger.

Though the rules of some of the clubs specifically prohibit gambling – for example, Rule 17 of one of the clubs which hold 'housey-housey sessions' states that "no gambling, drunkenness, bad language or other misconduct shall be permitted on the club premises" – there is gambling in all of them.

The largest sums are involved in what is called the 'Club Pool'. Each Sunday morning at the two largest clubs those members who wish to participate pay 1*s.* to have a disc bearing their membership number placed in a metal drum. The drum is revolved, and three of the discs are drawn at random. The owner of the first to be drawn wins (at one of these clubs) £30, the owner of the second wins £5, and the third £4.

Each Sunday morning Ashton's five ready-money betting offices are open, and bets are laid there with the bookmakers on the results of the club draw. Usually this takes the form of the member betting 1*s.* or 6*d.* that his own club number will be drawn first plus 1s. or 6d. that it will be drawn second or third. The book-maker agrees that if it is drawn first then he will pay £30 for the 1s. bet or £15 for the 6d. bet. If the member's number is drawn then, he receives not only the prize from the club, but also the winnings from the bookmaker. Occasionally a man who cannot afford the 1s. to have his club number included in the 'Club Pool' will lay a bet with the bookmaker that some other particular number will be drawn.

At another club the prizes are £22 for the first number drawn, £11 for the second, and £5 for the third. All the prize-money is derived from the shilling the members pay in order to enter the competition. A proportion of the sum collected in this way is set aside for the same sort of charitable object as part of the income resulting from the housey-housey receipts. It can be seen, therefore, that each Sunday morning approximately 1,500 of Ashton's 4,500 adult males participate in these lotteries.

Another form of gambling which takes place at all the Working Men's Clubs is that connected with the playing of cards, dominoes, and darts. The stakes are nearly always small, not usually more than 3d. or 6d.[1] Also each club has during the football season one or more 'sweeps'—each participant pays 3d. or 6d. for a team (he can of course 'buy' several teams) and the winner is the holder of the team which is first to score the requisite number of goals.

Each club also has its own 'bookie's runner'[2]—someone who is available at the club and with whom horse-racing bets can be placed. The 'runner' then takes the bets to the bookmakers for whom he is working, and receives a commission on the amount of money he has collected.

Equipment for other games, notably billiards and snooker, is provided at most of the clubs, and each game which is played has its handful of regular players. A small fee is charged for the use of equipment. It is very noticeable that today, as compared with the years of depression between the wars, games are not much played in the clubs. Between the wars the billiard tables, bagatelle boards,

---

[1] A few small groups gamble for high stakes (to the tune of losing several pounds in a night) at cards, but they are habitual and exceptional.
[2] The 'bookie's runners' are often disabled workmen who make a 'bit on the side' as they say, by being available at the club to take bets.

and skittles were in constant demand. Today bagatelle and skittles are never played. Though some of the clubs still have billiard and snooker teams competing in local leagues, at one of the largest clubs in Ashton the two billiard tables are stored in an upstairs room and not used.

There is little provision for 'intellectual' interests at the clubs if the 'intellectual' interest of talking is excluded. At one of the Working Men's Clubs visited in the course of Rowntree's second York survey[1] there was a stock of 2,684 books, and 200 members borrowed from the stock regularly. The club, with a total membership of 680, contributed £20 per year to the club library. In Ashton there are no libraries in the clubs. There are 'reading-rooms', but their sole provision is the national penny daily newspapers and several copies of the newspapers which specialize in horse-racing affairs. The second largest club in Ashton, with over 1,000 members in 1944, spent only £21 15s. 0d. on newspapers in that year.

Radio and television are also provided at the clubs (television is installed in only two of them). In just the same way as a high proportion of the newspapers are mainly concerned with sport so most of the radio and television programmes listened to and watched are concerned with sport. Indeed the radio and television sets are scarcely ever used for any other purpose in the clubs.

All the activities described so far in connexion with club life are open to all members *qua* club members. The club itself, however, is the matrix of subsidiary organizations whose members engage in activities which are not open to club members unless they are also members of the subsidiary organization. Each of these associations operating within the framework of the Working Men's Clubs has its own formal structure of an elected committee, an income of its own, and its own lotteries. Most of these organizations also organize their own celebration (in the case of sporting associations this celebration is an occasion for prize-giving).

Two typical examples of this kind of association are the Angling Club and the Tourist Club at one of the clubs. There are similar associations at all the Working Men's Clubs in Ashton.

[1] B. S. ROWNTREE, *Poverty and Progress*, 1941.

The angling club had thirty-six members in 1953, and each member wore a distinctive badge. The members vote annually by show of hands for a president, secretary, treasurer, and four committee men. They hold twenty club matches a year, and for these matches the members (the average attendance is thirty) travel by coach early on Sunday morning to suitable stretches of river in the neighbourhood of such places as York and Pickering–some 30–40 miles distant. The club members after fishing all day then spend the evening in some congenial public house before returning to Ashton. In 1953 the angling club prize-giving ceremony was followed by an entertainment given by a professional artist which developed into a series of 'turns' on the part of the members. The cost of the prizes and artist was borne by the angling club funds, which were mainly accumulated from the profits of lotteries held among the members.

Angling is an expensive sport. Each member of this club paid about £10 to £12 for his equipment. Then each time there is a club match each member pays about 6s. coach fare; and 3s. 6d. for tickets in the two lotteries which are held in the course of the day. It is also usual to take about £1 'for the pocket'–to pay for refreshment during the day. The members of the club are nearly all contract-workers at the colliery–the cost is prohibitive to day-wage men.

In the tourist club members are organized for the purpose of going on several minor excursions and one major tour in the course of the year. Approximately once a month the tourist secretary organizes a trip and hires coaches for the purpose. Usually about thirty club members and their wives decide to go.

It might be expected, and it is often stated, that the development of road transport has immensely widened the social horizon for the inhabitants of the once isolated mining communities. It is interesting to observe, however, that the people who go on the trips from Ashton clubs in fact seek out those activities in the places they visit which most closely resemble those provided in Ashton. Thus a high proportion of the journeys of this particular tourist club were to Working Men's Clubs in other towns. They seldom go farther afield than places which are within a 25-mile radius.

The same consideration applies to the major excursion of the year, when about fifty members of the tourist club go to Wembley to watch the Rugby League Cup Final. (It must be remembered that reference is being made to the tourist club of only one of the six Working Men's Clubs. There are similar clubs at the others.)

In 1953, forty-five members of their tourist club visited London (Wembley). The capacity which they showed in selecting from the London environment traits which resembled Ashton was remarkable. And having found, for instance, the Ashton-like public houses instead of avoiding them, they insisted on spending the evening in them. It is not being suggested that the Ashton miner is peculiar in this respect. He is able to withstand the influence of a new environment by selecting only those aspects of it which fit in with his established pattern of existence–a pattern which is a product of the environment of Ashton–and attempting to ignore the rest. The so-called 'widening of the horizon' therefore means far less in the lives of actual miners than an objective enumeration of the new possibilities would seem to indicate.

The tourist club is different from the angling club in many respects. Whereas the latter retains approximately the same membership year after year, the membership of the former changed almost completely in the course of the single year 1953. At each 'meeting' (i.e. at each tour) four or five who were present at the last meeting were not present, and their places had been taken by others. Also, the membership of the tourist club was much more heterogenous than that of the angling club. All grades of Ashton miners and workmen were represented, and women were allowed to accompany their menfolk on the tours.

In other respects the two clubs resemble one another. The tourist club has its elected committee of president, secretary, treasurer, and two trustees. It raises funds by holding lotteries, and distributing only part of the money wagered. It also holds its own annual dinner: at the 1953 annual dinner there were ninety members and friends present.

It can be seen that the clubs are vigorous centres of 'social intercourse', but are hardly active in 'mental and moral improvement'. With regard to 'mutual helpfulness' the clubs' achievements in this respect are not at all impressive. This is apparent both at the level of the formal mutual helpfulness organized by the club, and at the level of informal aid between the members.

In essence the Working Men's Club is a co-operative society for the purchase and sale of beer. The elected officers of the club purchase beer, spirits, tobacco, etc., on behalf of the members. These commodities are then sold to the members at the ordinary commercial price. This 'profit' from these sales is then available for the benefit of the members – it is in fact a 'mutual helpfulness' fund. The main use to which the funds are put is that of periodically providing members with their free beer or beer below cost price. Thus it is customary for each member to receive, say, 8 pints of free beer during each of the following holidays: Christmas, Easter, Whitsuntide, and August Bank holiday. The second use to which the profits are put is in providing the kind of concerts which have been described earlier. Thirdly, the profits are used in providing games equipment, newspapers, and when necessary, such things as a television set. The members of the club who do not spend a lot of money at the club are in that way 'helped' by those who do. There is no suggestion of 'mutual helpfulness' on the basis of need.

The Working Men's Clubs can therefore be seen to reflect in their behaviour as organizations the thriftlessness of their members. The clubs' funds are spent as quickly as they are accumulated, just as the members' wages are spent as quickly as they are earned. To take one example: an Ashton club with a membership of over 1,000 and an income of just under £15,000 in 1944 possessed a total investment income of only £15 from Defence Bonds. The result is that in the event of industrial depression the clubs have always been forced to borrow to meet their necessary costs. In 1926, for instance, Ashton's largest club borrowed £1,000 from a large brewery company.

The funds are, however, used to a slight extent to help needy members, or members who are presumed to be in need – namely the old age pensioners. Every year each club gives the old age pensioner members one or two 'treats' – either in midsummer or at Christmas or both. The old men are entertained to a meal free of charge, to free drinks, and each is usually given about 10s. pocket money. There are also 'treats' for the members' children, and sometimes a party at Christmas. The cost of these 'treats' is borne by the income from lotteries organized for the purpose, supplemented by a subvention from the general funds of the club.

Just as the formal mutual helpfulness of the club mainly consists of free or cheap beer, so the informal aid between club members mainly consists of one member buying drinks for another member who is *temporarily* unable to afford them. It is expected in each case, however, that generosity will be reciprocated when the positions are reversed. When a member is permanently impoverished he does not expect to be bought many drinks, and usually keenly feels that if he is bought some drinks he should repay in some way. An old miner who was permanently disabled at the colliery as the result of an injury to his leg had this to say:

"When I was lamed 5 years ago, and I came out of hospital on crutches, I. J. was the first one to buy me a drink. Since then he has given me 2s. 6d. or 3s. or whatever change he has happened to have in his pocket, so that I could have a quiet drink.

"I can't pay him back straight away. But I had a pair of glasses and I. J. tried them on. 'Champion! Just right!' he said.

"'Put them in your pocket,' I said.

"'What do you want for them?' he said (i.e. how much do you want to sell them for?).

"'I want you to put them in your pocket, that's what I want! My eyes are too far gone for reading glasses.'"

Public houses

Having discussed in some detail the part of the Working Men's Clubs in Ashton's life, it is not necessary to say much about the public houses, since in Ashton the two institutions play much the same part. They cater for the same type of clientele. Though women are not prohibited from entering the public houses during the week, in fact it is only at week-ends in most cases that they are to be found there. In practice there is no difference between the clubs and public houses in this respect. Just as there are the

exceptional clubs which have women members, there are the
exceptional public houses where a few women[1] are regular
customers during the week.

The public houses, in fact, broadly duplicate the facilities of the
clubs. Thus there are six Working Men's Clubs with an annual
turnover ranging from about £25,000 for the largest to about
£6,000 for the smallest. Similarly there are seven public houses
with a similar range of turnover – from about £30,000 in the case
of the largest to about £6,000 in the case of the smallest. Though
the public houses have neither concerts[2] nor 'Club Pools' there is
otherwise little difference between their activities. The main
interest is talking over a pint of beer. There is the same division
between the 'work and sport' conversationalists and the 'Best
Room' people who have a wider outlook. The games, too, are the
same – dominoes, cards, and billiards or snooker. Associated with
each public house are the same kinds of subsidiary organizations
strictly analogous in function to the Working Men's Club
organizations. There are in addition organizations associated with
the public house which are branches of national organizations but
which in fact take the form of small social clubs crystallized out of
the general clientele of the public house. Examples of these are the
Ashton Branch of the British Legion and the Royal and Antediluvian Order of Buffaloes. Nor do the two institutions differ in the
manner in which they cater for the children of customers, and
customers who are old age pensioners. By means of regular lotteries
many outings are arranged from money raised for 'Children's
Outing Funds' to give the children of the customers who participate in the lotteries a day at the seaside.

Both these institutions – club and public house – which are so
very similar, continue to exist separately and retain their individuality. This is possible because each has advantages that the other
lacks. The profits of the clubs are used to provide the club members with cheap or free beer regularly. Of course this does not
happen at public houses. On the other hand the publican is more
sensitive to the wishes of his clientele than are the club committee

---

[1] Only *old women* go to the public house during the week unaccompanied.
[2] Some of the public houses in the neighbourhood of Ashton though not in Ashton itself,
do hold free concerts.

men to the wishes of their members, because the publican's livelihood depends upon his relations with his customers.

Sport

There are only two other leisure activities which regularly bring large numbers of Ashton people together. They are sport and the cinema.

Those activities connected with sport in Ashton resemble the club and public house in being primarily the domain of the male. The most important of the 'sporting' activities is that of supporting the Ashton Rugby League team. The term 'supporting' has been used advisedly. It is meant to signify that the activity is not a mere passive process of watching and taking pleasure in the display of a particular skill. Each game is an occasion on which a high proportion of Ashton's males come together and participate in the efforts of Ashton to assert its superiority (through its representatives) over some other town (through their representatives). The Lynds expressed the same view in connexion with the part the basketball team played in the life of Middletown. The team's victory, the Lynds said, gave each supporter the sense of being "a citizen of no mean city, and, presumably, no mean citizen".[1]

In 1953 the following attendances at the home matches of the Ashton team were recorded. It must be pointed out, however, that a variable proportion of these attendances was accounted for by supporters of the opposing team and supporters of the Ashton team living in places near to Ashton. The magnitude of these attendances can be placed against the fact that the total adult population (male) of Ashton – i.e. males aged 18 and over – is a little over 4,800.

That the interest in the game is not based upon the desire to witness a display of skill was well demonstrated in November 1953. In that month the whole of the international Rugby League match between England and France was televised. In the television room of one of the clubs there were never more than nine men watching the match. There were 3,400 at the match in which Ashton was playing at home.

[1] R. S. and H. M. LYND, *Middletown*, New York, 1930, pp. 486–7.

TABLE IX

Attendance at the matches in which Ashton Rugby League
team played at Ashton in 1953 (i.e. the latter half of the
season 1952–3 and the first half of the season 1953–4).

Home match	Attendance
1	3,300
2	2,800
3	4,000
4	4,200
5	5,000
6	10,000
7	3,500
8	3,200
9	6,100
10	7,700
11	7,800
12	3,500
13	3,500
14	4,500
15	3,700
16	3,400
17	4,400
18	3,500
19	3,800

With regard to the importance[1] which the defeat or victory in
these matches assumes for the supporters, it is a joke in Ashton that
when the Ashton team is defeated "two thousand teas are thrown
at t' back o' t' fire". The men are said to be too distressed to eat.
A miner said of his colleagues in the concert-room of a club one
Saturday evening when Ashton had defeated a leading team in the
Rugby League: "We beat . . . today. Look at everybody! They
would still have been happy tonight if the worst turn in the world
had been on the stage!"

It is interesting to note that the general body of the supporters
of the Ashton team has within it various subsidiary organizations.
These are the 'Supporters' Clubs', of which there were thirteen in

[1] Because of the labour troubles in the Beeston seam of Ashton Colliery in 1953 affecting
production (see Chapter III, 'Trade Unionism in Ashton') no comparison between the
victories and defeats of the Ashton Rugby League team and the figures of coal production
was possible.

Ashton in 1953. In addition to their primary interest in raising funds for the football club (they raised £425 in the season 1952–3) the members of the supporters' clubs engage in purely social activities – i.e. they organize parties and outings. They also 'treat' the children in the same ways as do the public houses and clubs. In 1953 there was a 'Junior Supporters' Outing' for 300 children. The general body of supporters, too, engage as supporters in activities not directly connected with football. Several 'Rugby League' activities, such as ballroom dances are held in the town baths in the course of the year. At one of these dances an Ashton Rugby League Queen is selected by a panel composed of the town dignitaries. Later she will compete with 'Rugby League Queens' from other towns, in an attempt to show that Ashton possesses beautiful women as well as strong and skilful men.

No other sport can be compared with Rugby League in its importance for the general life of the community. Many other sports are played in Ashton, but they are for the most part the concern of only those who play them. Each sport or group of sports has its own club organized around it. There is a cricket, tennis, and a water polo club, and many bowls, billiard, and table tennis teams. All enter into competitions with similar teams from other areas. Also each club or team, organized as it is for the purpose of pursuing some particular sporting activity or group of activities, also organizes convivial outings, annual dinners, and prize-giving ceremonies for its members.

Betting

There are two forms of sport which while they find neither players nor supporters in Ashton, yet assume a considerable importance in one respect. The sports are Association Football and horse-racing. Their importance consists in their use as media for betting.

The descriptions of Ashton leisure activities given so far show that gambling is not confined to football pools and horse-racing. It dominates almost every form of leisure activity. The primary reason put forward for this domination has been discussed above in connexion with the general influences which determine the pattern of social life in Ashton. It is that the miner, because of the various forms of insecurity which beset him, cannot hope to escape the limitations of the miner's existence by saving. He can only escape the heavy, dirty and dangerous work by 'luck' in a big way.

A second reason for the popularity of gambling is that the increase in miners' earnings in the last few years has given him a large margin of income available for free spending. It has been already explained that the miner prefers to have a large proportion of 'free income'–income, that is, which is left after what are regarded as being necessities have been purchased. He is liable to be deprived of this margin due to injury or demotion, and therefore avoids becoming too dependent upon it. Gambling is under these circumstances a very attractive way of spending money and the miner becomes an easy prey for those who wish to engage in what the 1933 Royal Commission on Gambling called "the mass exploitation for private financial gain of the . . . propensity to gamble".[1]

Other reasons which appear to give gambling its appeal–and these apply to gamblers in all occupations, not only mining–are the desire for self-assertion and the satisfaction derived from being 'lucky'. The satisfaction derived from 'assertion' and 'luck' are logically speaking contradictory, yet it is clear from observation that many people who gamble in fact derive satisfaction from both sources. The desire for self-assertion finds satisfaction in the belief that someone with knowledge of the intricacies of the sport has a better chance than someone who lacks first-hand knowledge. A winner feels that his superior skill has received its just reward. The belief in luck, on the other hand, ignores the question of merit, and the winner feels that his success is in some sense proof that he is a favoured son of Providence.

Of the two main types of gambling (football pools and horse-racing) which are not subsidiary to membership of some club, or indulged in merely to add spice to a game of cards or dominoes, football pools is the most widespread. During the football season in an average week 6,000–7,000 football coupons are delivered in Ashton. There are very few adult miners who do not participate

[1] *Report of Royal Commission on Gambling*, 1933, para. 219.

in this form of gambling. Many who take part, however, do not consider themselves to be gambling at all. The reason for this is that the chances of winning a considerable sum are very slight— though very few participants realize just how slight they are. There is not therefore that feeling sometimes found among people who bet on the results of horse-racing–a confidence in their ability to back a succession of 'certainties' and thus solve all their problems. For most Ashton people who fill in football coupons the feeling is rather that "I am just giving myself the chance, if it was to come to me." They find it hard to understand why anyone should throw away the possible chance–however remote–of £75,000 just for the sake of 1s. Generally this practice involves an hour or so one evening a week during the football season, and an outlay of a few shillings–"just so as we have a chance".[1]

The second type of gambling is that which is centred on the bookmakers' offices in Ashton, of which there are five. The smallest of them is in a hut attached to the proprietor's private house. The other four are all dingy premises on Ashton's main roads. They have ordinary shop fronts, the windows of which have been painted. Inside each of the shops the scene is bleak and dusty. The floor is either composed of the bare floorboards, or covered with worn and broken brown linoleum. The furniture consists of a few chairs, a wooden form and an old table. A large blackboard covers one wall on which the details of the afternoon's races are recorded –the time of the race, the runners, and the odds which are being offered by the bookmakers on the course. On another wall are copies of the day's sporting newspapers. There is a loud-speaker– the 'blower'–which gives up-to-date information direct from the race-courses, including commentaries on the races. There is finally a counter–in some cases there is a wire partition between the customer and the bookmaker. The most common method of placing a wager is for the customer to hand across the counter a slip of paper, telling the bookmaker (1) the name of the horse he

---

[1] The following conversation was overheard:

A. "I just missed the Treble Chance last week, just one match let me down."

B. "I don't know why you bother at all. I never do. You've no chance at all."

C. "I've a better chance than you, at any rate, if you don't bloody well send them in at all."

wishes to back, (2) whether he wishes to bet only that it will win
or to place an additional bet that it will come second or third,
and (3) whether he wants to bet at the odds as they stand at the
time he is making the bet, or as they stand at the start of the race.
This slip of paper is signed by the customer, usually with a
pseudonym which is often a 'lucky' name or number for the
customer. With the slip of paper is included the money which is
being wagered.

Although regular gambling on the results of horse-races is not as
widespread as gambling on the results of a selection of football
matches, it is probably more important in the life of Ashton. The
bookmakers have a weekly turnover which resembles in magnitude
that of the Working Men's Clubs. Thus the turnover of one of the
bookmaker's shops in Ashton is known to be in the region of £700
per week in the 'flat season'. The separate sums which comprise
this total is composed in the great majority of cases of bets under
£1. Usually the bets are from 2s. to 10s. though one man may take
several such bets in the course of the afternoon. This £700 turnover does not, of course, represent £700 in miners' earnings per
week. The bookmaker pays his running costs, takes his profit, and
returns the remainder of the money to his customers in the form
of winnings. These winnings, in so far as they are used to make bets
with the bookmaker thus reappear in the weekly turnover again
and again. The weekly turnover is, therefore, misleading if it is
taken as a true measure of the *cost* of gambling to Ashton. It is,
however, if considered together with the size of the individual bets,
and the number of customers, a true measure of the *extent* and
intensity of gambling.[1]

Most of the people who use the bookmakers' shops simply
walk in, place their bets and walk out again. For them the place
of the bookmaker's shop in their lives is like that of the football
pools – an hour's thought and a few shillings a week. Many of their
bets indeed are the same as those with the football pools, in that
the chance of winning is small but in the event of winning there
is a prize which is large in relation to the stake. They bet that the

---

[1] The profits from bookmaking in Ashton are difficult to estimate, but they are certainly
more than £7,000 per annum.

'outsiders' will win–horses considered to have so slight a chance of winning that knowledgeable bookmakers are willing to offer long odds against them. The view of this type of gambler was put by one of them in the following terms: "What is the use of a working-man risking £1 to win £1 ? Now risking £1 to win £100–that's different !"

Though most of the individuals who frequent the bookmakers use them like they use other shops–they *purchase* their chance of winning a certain amount of money and leave the shop–the great majority of bets are wagered by the 'regulars'. Each of the four main bookmakers in Ashton has his forty or so regulars and these constitute the core of the gambling community. Only about five or six of these are women. The male regulars use the shop as a kind of club. The women, even if they are regulars, never use it in this way. They place their bets and leave. When one of the regulars is on afternoon shift he may spend the morning in the shop–that is even when there is no racing–with others like himself.[1]

Cinema-going

Clearly the leisure activities so far described bear out the generalization with which this discussion commenced. Life in Ashton, if it is to be judged from these activities, is undoubtedly vigorous and frivolous, in the sense given to the term frivolous in this context. These activities have also been shown to be predominantly the concern of the males.

When the cinema, the only other activity which brings Ashton people together regularly in large numbers is considered, the position is seen to be somewhat different.[2]

In the first place, women are admitted on an equal footing with men, and audiences are in fact composed of roughly equal proportions of the two sexes. Women preponderate when certain types of film are shown–for example, musical romances–and

[1] N.B. Our concern has not been to give an interpretation of that very complex series of social phenomena coming under the heading of gambling, but only briefly to describe gambling in Ashton in a manner consistent with our general account of social relations and ideology. It is hoped to publish later a more complete account and analysis of gambling in Ashton.

[2] The Ashton cinema has a seating capacity of more than 800. Average weekly attendance about 4,000.

men when certain other types are shown such as war pictures and 'westerns'. It is not at all unusual for women to attend the cinema unaccompanied. For many women with young children the cinema is the sole relaxation outside the home and they often come alone while the husband looks after the family.

Secondly even large attendances at the cinema are not evidence of 'vigorous social life'. The relationship subsists not between the different members of the audience but between each separate member of the audience and the film being shown. J. P. Mayer states in his *Sociology of the Film*[1] that "the atomized type of cinema-goer and the autark pair form the majority". The majority of people who attend any sort of club or meeting *go* singly or in pairs. In the course of the meeting, however, there is a coming together of the individuals and pairs – they cease (to continue to use Mayer's terms) to be atomistic and autarkic. The account of the Working Men's Clubs Saturday concert given above describe this happening. In the course of a cinema performance on the other hand there is no such coming together. Even the relationships between the people who attend the cinema together are virtually suspended during the performance.

The cinema differs from other Ashton leisure institutions in that it is not predominantly a male preserve, and it is not vigorously 'social'. What the audience at the Ashton cinema requires is above all – action. The most popular types of film are therefore the coloured 'western', the spectacular adventure story, and the slapstick comedy. Pictures about the 1939–45 war are also popular. Musical romances appeal to the women. All these films resemble one another in their irrelevance to the problems of the day – they are escapist in that sense – most of them quite blatantly and unashamedly so.

Similar considerations apply to the influence of the cinema as were seen to apply to coach trips to places outside of Ashton. There is in both cases a selection from the environment of those aspects which least disturb the Ashton pattern of life. When Ashton people visit other places they protect themselves from alien influences by searching out those activities which most resemble the

[1] London, 1948.

familiar. When they visit the cinema they protect themselves by attending only those films which portray events so utterly remote from any they know that their portrayal has no real impact on their lives. An exception to this is that among young people such things as personal appearance, clothing and make-up, are affected. In this respect, the influence of the cinema is reinforced by that of commercial advertisements.

Minor leisure activities

In addition to these activities which large numbers of Ashton people regularly engage in, there are some activities in which many engage occasionally. There is, for example, the annual 'Gala Day' organized by the Urban District Council. The programme for 1954 included the following items:

> Motor Cycle football match.
> Spink and Spink–Aerial Gymnasts.
> Billy the Kid and Partner–Wire Rope Walkers.
> Donkey Derby Race Meeting.
> Grand Fireworks Display.
> Fun Fair.
> Bar.

Much less popular is the annual Miners' Gala for the whole of the Yorkshire Area of the National Union of Mineworkers which is held at various places in Yorkshire. On the whole the Yorkshire Gala has never attained the popularity of its equivalent in Co. Durham.

There are also numerous organizations which cater for smaller groups of people in Ashton who share a particular interest. In Ashton these voluntary organizations range down from the Working Men's Clubs at one extreme, with their congeries of subsidiary organizations, to the street organizations constituted to provide the children with an annual trip to the seaside.[1] Several such organizations have been described above.

Of those which have not been discussed the most active is the Old Age Pensioners' Association. In Ashton Urban District there

[1] Many Ashton streets have "Children's Outing Funds" to provide the children of the street and incidentally often its old people, with a summer outing or Christmas party.

are 1126 people aged 65 and over. There are about 100 members of the association, and the average attendance at meetings is about 80.

It was suggested earlier in the chapter that the explanation of the types of leisure activities existing in Ashton, was to be found in the circumstances of the miner's life. It was said that two attitudes of mind existed–the 'frivolous' and the 'serious'. The social behaviour resulting from the first attitude of mind has been described above. Behaviour resulting from the 'serious' has to be considered.

The Labour Party

One association which can be regarded as being transitional between one type of behaviour and the other is the local Labour Party. Here can be found some traces of what was called earlier "a serious concern with the miners' predicament, and a determination to seek out its causes and its cure". Among the male members only between fifteen and twenty take an active part. All the other male members–nearly all Ashton miners are members of the local Labour Party through the affiliation of the trade union–take no part whatsoever. This score or so is composed of those ten public-spirited men who are town councillors, and other public-spirited men who hope to become town councillors. During 1953, four general meetings were held. At each of them serious matters were mentioned, but they always assumed less importance in the discussions than did other subjects discussed. This can be seen by looking at this account of a local Labour Party general meeting in September 1953.

"Twenty-two members were present, of whom thirteen were women and nine men. The men knew all the fine points of procedure, and made much of them. The women seemed to be ignorant of procedure and all their contributions at the meeting were treated with amused tolerance by the men.

"The minutes of the previous meeting were read and passed as a true record. Among the correspondence was an invitation to the local Labour Party to send delegates to a British Asian Fellowship conference at the county town. There were no volunteers.

"'I've a letter here about British Guiana,' the Secretary said, 'It's a bit of a long letter. Does anybody want me to read it?'

" 'Don't bother,' one of the ladies cried out. 'We've enough trouble in Ashton.' The letter was accordingly left unread.

"The Secretary then drew the meeting's attention to some literature which had been forwarded to him. 'But I'll just let it lie on the table. We wouldn't have any financial funds left if we bought everything that was sent.'

"Then followed a lengthy discussion about the submission to the Divisional Labour Party of the names of prospective J.P.'s. Its general tenor was that Ashton was badly neglected in this respect, that there were 'dark plots' and 'as much as secret service as was ever known' working against any Ashton person becoming a Justice of the Peace. What was more 'the more we talk about it, the more secretive it becomes'. The discussion was monopolized by the 5 men whose names had been submitted.

"For the remainder of the time the meeting discussed the arrangements which the 'social' sub-committee had made for the annual dinner and dance."

The women have their own "Women's Section of the Ashton Local Labour Party" and there are nearly 100 members. It is one of the largest women's sections in Yorkshire. There are fortnightly meetings attended by between twenty and forty members. All are middle-aged or old. At these meetings they hear speakers on such subjects as 'food' (the problems of a coming world food shortage) and "Will there be an election in the Autumn?" The growth of this women's section in the years after 1950 explains the large numbers of women at Labour Party functions. However, the interest and activities of the women in the Labour Party in Ashton are 'social' rather than political in character and in addition are often typically 'feminine' in Ashton terms.

This predominance of social rather than political interest in the Party is however not restricted to women members. For example, while there were only 26 members at the annual general meeting of the local Labour Party in 1954, there were 153 at the annual dinner. Also, the core of the women's section–those members who are the most regular in their attendance–are the members of the Ashton Labour Ladies Choir. This choir performs at Ashton meetings of all kinds.

In Ashton there are no other important ways of spending leisure time in association with one's fellows. It is necessary to mention, however, that there are facilities for adult education which, as might be expected from the foregoing account, are little used.

NON-INSTITUTIONALIZED LEISURE PURSUITS

One other category of leisure remains for consideration – that which is primarily individual or familial. The description of Ashton given in the opening chapter made it clear that no one in Ashton is far away from the countryside. There is therefore ample land within easy reach of Ashton – indeed, at the end of each street – suitable for allotments. Many Ashton families do in fact cultivate allotments. In 1953 there were 3,837 private households in Ashton; 1,100 of these had allotments covering a total of 119 acres. There are in addition the gardens of the council houses, and of a few of the non-council houses. Among men who have allotments there are those, a minority, who spend so much time there as to separate themselves from their families for considerable periods.[1]

More strictly within the circles of the family, the leisure activity of reading is carried on. Nearly all the books which are read are borrowed from the public library, an amenity of comparative recent standing, for Ashton did not possess a public library until 1925. The existing library building was not erected until 1934 and then the total stock of books was 2,300. Ashton's first full-time librarian was not appointed until 1942. In 1953 the book stock stood at just under 7,500 of which 2,700 were non-fiction books, 3,700 were adult fiction and 1,100 were junior fiction. There were 2,188 registered borrowers in March 1953, and of these approximately 1,500 used the library regularly, the monthly issue of books fluctuating between 5,000 and 6,500. Reading must therefore be ranked with the clubs, the public houses, sport and gambling in respect of the number of people concerned.

Only one-quarter of the library books issued were non-fiction. In general it can be said that Ashton's reading serves the same purpose as its cinema going. That purpose is to convey the reader into a world quite different from his own or her own, and remote therefore from his or her problems.

[1] See Chapter V, 'The Family'.

Among books read by the men, 'westerns' are markedly the most popular. Partly the appeal of this type of book is that of all escapist literature, partly also their appeal is based on the close adherence to a formula, so that the reader feels at home, so to speak, in the strange country he has entered. Many readers are so reluctant to lose that feeling of being on familiar territory that they specialize in the novels of a particular writer. The desire for the 'formula' is perhaps further shown by the popularity of authors who use the device of writing a series of books about the exploits of a particular character.

The only other type of book which is demanded by Ashton males in large numbers is the so-called detective-thriller. In Ashton the book which is more of the 'thriller' type – i.e. which depends upon the action of the plot – is preferred to the 'detective' type of novel – the novel which depends upon the problems to be unravelled.

In Ashton women are predominantly the readers of the light romantic novel. There is some demand for 'westerns' but this is partly because in recent years the 'western' has tended to become also a 'romance'. The light romantic novels, however, are rarely read by the men. Among those authors whose works unfailingly appeal to Ashton women are Kathleen Norris, Margaret Pedlar, Simon Dare, Bertha Ruck, and Ruby M. Ayres.

The other main leisure activity centred on the home, the radio,[1] has not proved a sufficient counter-attraction to the leisure facilities provided outside the home. Listening to the radio is much more a woman's than a man's pastime in Ashton. The word 'listening' in this context can, however, be misleading. It is common to find in Ashton households that the wireless is put on in the morning and remains on until there is a programme being broadcast which is disliked. If suitable programmes cannot be found as an alternative then the radio is switched off. Selection is then a negative process of switching off what is not liked rather than the positive process of tuning in to a programme which is desired. The radio accordingly tends to provide a background to other activities rather than

---

[1] At the time of this study T.V. was something of a novelty in Ashton. It was thought that a study of its effects after so short a time would be of little value.

to be the basis of a separate activity. In this Ashton does not, of course, differ from most other areas.

THE CHURCHES

In Ashton there are two Anglican churches, one Anglican Mission, eight Nonconformist churches and chapels, and one Roman Catholic church. These undoubtedly played a far more important part in the life of the community in the past than they play at present. Thus the largest Methodist church was constructed shortly before the 1914–18 war. The congregation–at that time adhering to the Wesleyan connexion of the Methodist church– originally met elsewhere, but found the premises too small. The new church had a capacity of well over 600, and until about 1926 according to a leading member of the church there were very few vacant places at services. The membership of the men's Bible Class alone in the days before 1926 was 80. Today the total membership of the church is 80, and men are in a minority. The average Sunday attendance in 1953 was 8 women, 4 men and 12 children at the morning service, and 22 women and 10 men at the service in the evening. One of the smaller Methodist churches–formerly a Methodist Free Church–has retained rather a higher proportion of its members. In 1912 the membership was 56, in 1919 it was still 56 but by 1930 it had fallen to 48 at which figure it remained until just before the 1939–45 war. After the war there was a further decline and by 1953 there were only 33 members.

The church services registers of one of the Anglican churches and of its mission indicate that an average of 100 Ashton people attend Holy Communion every week. The miners and their wives are not represented within this group in proportion to their numbers in the town, however, and the Ashton tradespeople are over-represented.

The churches have their subsidiary organizations in the same way as the Working Men's Clubs. Like all Ashton organizations these cater for the social interests of their members. Their aim, however, is invariably the moral, spiritual and intellectual betterment of their members, and in this way they differ radically from the Working Men's Clubs organizations.

Women's leisure activities in general (apart from cinema-going and week-end visits to the club and public house) are not so rigorously separate from work as they are for the men. The radio is a constant background of recreation while the women are working in their houses. Work is suspended intermittently when neighbours and friends come 'callin'', come that is, to gossip. In the time spent on it, 'callin'' is the main leisure activity for women in Ashton.[1]

There is evidence to suggest that leisure time is spent in Ashton today in much the same ways as it has been spent since the nineteen-twenties. (Before then the churches exerted a very much stronger influence on the way leisure time was spent.) The *mechanism* by which change is minimized is that of avoidance. Those agencies which would tend to induce change – for instance, the cinema, road transport, and literature are neutralized to a large extent in the way which has been shown.

Furthermore the evidence suggests that the home in Ashton is considerably less important than places outside the home as a centre of leisure activity. And as the leisure facilities outside the home are for males rather than females, the husband and wife move for the most part in different spheres. The results of that separation belong however to a detailed discussion of the family in Ashton.

[1] One case was noted of a 26-year-old wife whose husband and two young children left Ashton to live in Slough. Within the year the wife's dissatisfaction with Slough forced the husband to relinquish his job and return to Ashton. "I was used to callin' with neighbours," she said, "they don't do it down there ! I couldn't get used to the secluded life." Callin' is also a favourite pastime when women are out shopping.

'Callin' '–'Call' ('a' pronounced as in 'shall')–to gossip.

CHAPTER V

The Family

THE FAMILY IN ASHTON

THE family is a social formation which may be studied from various view-points. A detailed analysis over a period of the types of interaction between its constituent members would be valuable. Another approach, fashionable among some anthropologists, is to discuss family organization mainly in terms of its contribution to building up the type of personality fitted to certain roles and attitudes in the society into which children must grow. This is the process known as socialization. Our concern is not with the details which would form the principal material of either approach. In this section the aim is to present a description of the family in its relation to the structure of working and other institutions in the same community. An attempt is made to show that the basic features of family structure and family life derive their character from the large-scale framework of Ashton's social relations. This framework consists of Ashton's basic industry and the relations which it enjoins on Ashton's population, together with the institutional life which has grown upon that basis. Previous chapters have outlined this total social system; we now turn to the place of the family within it.

It is possible to speak of 'the family in Ashton' even though the detailed life of no two families is the same. One couple may be younger than thirty-five, have an only child alive and live in a modern council house a mile from the colliery, while another may be almost sixty, have a colliery house built in 1893 on the edge of the pit-heap, and share the house with sons and daughters and their families. At those limits significant differences do emerge, for all sorts of reasons, but these differences, like the differences noted earlier between certain grades of miners, exist within a basic framework of similarity. All miners' families are dependent for their livelihood on the local colliery. From the wages paid for the work at this colliery over 60% of Ashton families depend for their sustenance, and many of the remainder depend on ancillary occupations which the colliery originated and maintains. More than four-fifths of the families at present living in Ashton originated in that period of 1895–1908 when the community truly developed from a village to an industrial town based on the collieries at Ashton and Vale. These families have given individuality and continuity to a community, Ashton, to which their attachment is maintained through the continued existence of the primary cause of their coming together, employment (with little immediate alternative) at the local collieries.

Most of the Ashton families, therefore, have a continuous record of existence in the town for fifty to sixty years at least. As a consequence the ties of any particular individual or family to the locality are strong and manifold if compared with many of the families living and working in larger industrial towns. Since the economic depression beginning in the late 1920's, some 7% of Ashton's population has left, thus weakening some of the ties of the community, but not dealing nearly so heavy a blow at community life as larger-scale urban developments in the larger towns. The building of very large council housing estates in a city, drawing their inhabitants from a waiting-list compiled from a population of say, half a million, or even a hundred thousand, will break up those long-standing communities still existing in working-class areas. It does this without providing those stimuli for new communities which built up the old ones, i.e. a common workplace and a limited range of activity including most of the local people. Now in towns the size of Ashton, the building of a new estate, and even the drift of 1,000 inhabitants, will not tend to break up the community. The range of activities available to the inhabitants of the new houses is still concentrated in Ashton itself, the social contacts of each family, the proximity of known relations and friends, are only slightly and insignificantly modified. Ashton has not reached a size where such changes can rupture the system of face-to-face relations.

The nearby town of Calderford (population 40,000) is an

interesting study of the limits at which the factor of size works against the influence of disintegrating factors in community life. In a town of Calderford's size, face-to-face relations between all the inhabitants are, of course, impossible–even Ashton does not quite achieve this–yet usually every man or woman in Calderford can trace some sort of acquaintance with a person of his or her own generation. They will have shared directly or indirectly some contact at the various places of employment available in and near the town or in the several places of social activity and entertainment in the centre of Calderford. Now Calderford came into being as half-a-dozen colliery villages more or less grew into each other in the years of rapid mining development (1890–1910), and the conglomeration of shops, clubs, public houses and later cinemas and dance-halls in the new 'town centre' provided a focus for all the villagers. The latest housing estate schemes, however, manifest a growth outwards from the borders of the town. New estates are being built on the outskirts of a town which was just small enough and sufficiently closely built to maintain some sort of community life. Ashton is still so small that geographical location within it is comparatively insignificant in affecting the existence and nature of the social ties between families. In Calderford, however, the movement of families to these outer limits of a town of 40,000 is already breaking up the long-lived and more intimate unity of the town which existed in the rather ugly and crowded, often broken down, housing conditions of earlier days.

In Ashton each family is born, grows, and gives birth to other families or branches in a set of ties inherited from the recent past of 50–60 years, a past which spans most living memories. One other factor besides this set of shared relations, in time, and space, with other individuals and families, acts as a factor for keeping families in one place. Mining is an occupation with few attractive prospects to the outsider. The colliery village is never attractive, to say the least. Is the simple fact of inertia based on custom and adaptation to a given place and set of social relations with known people sufficient to explain the hold of a place like Ashton? It is suggested that the additional factor is the existence and persistence in a mining community of its own standards, in particular its basic living standards limited by the weekly wage. In actual fact this 'additional factor' is only a part of what we have called the persisting set of common social relations, for it is a relation held in common by miners and their families with the work available in the area, its reward, and the life made accessible through that reward of labour. Yet it is the one relation which is sufficiently all-embracing and basic to determine in large measure the scheme of the whole set of social relations. This basis is no different from the common fate of wage-workers which we have discussed earlier; it is significant for a discussion of the family, as well as for problems of industry, because the family is normally a group of father, mother and children all dependent on the wage of the father. This is especially true of a mining community like Ashton, where facilities for the employment of women have always been slight. The pure economic fact of man's being the breadwinner for his family is reinforced by the custom of family life, the division of responsibility and duties in the household, and the growth of an institutional life[1] and an ideology which accentuate the confinement of the mother to the home. 'Woman's place is in the home' is a very definite and firm principle of thought and action in Ashton. Here we see already that, given the basic fact of a certain economic and social framework, family life and the accepted division between the sexes can build up a set of mores and ideas with an intrinsic force in daily behaviour.

The building-up over time of a commonly accepted standard of what sort of life, particularly in terms of its material prerequisites, a family leads, is an extension of that principle which holds miners to their work.[2] It was impossible to answer the question "Why does the miner stay in mining?" without seeing that he had grown into a framework of social relations in the industry which by various devices enabled him to maintain that level of comparatively high wages which a miner expects. The question "Why do mining families stay in mining villages?" is answered satisfactorily only in a similar manner. In addition to the facts of the ties of community life it must be seen that mining affords a level of wages

[1] See Chapter IV, 'Leisure'.   [2] See Chapter II, 'The Miner at Work'.

traditionally higher than in other industries for men without long-term education or technical training. This is the level of wages on which the miner knows his family can maintain the standards of comfort and expenditure in terms of good food, a warm fire, and the capacity to enjoy leisure hours and festive occasions with that freedom of spending for which miners are well-reputed. Comparatively high wages in mining are by no means a post-war or post-1939 phenomenon. A sufficient labour force could only be recruited to such an industry if the margin of wages between mining and other occupations for workers of the same education and skill were sufficient to attract it. Even in post-war conditions, with nationalization, security of employment and a higher level of wages than other industries, it has hardly been possible to maintain the labour force at the level required for national security. In 1954 the general secretary of the National Union of Mineworkers, clashed with the chairman of the National Coal Board in urging that output could not be raised until wages and conditions were improved so as to facilitate recruiting to the industry.[1] In the inter-war years of depression and under-employment, when the standard was not maintained, miners did *not* all stay in mining, as witnessed by the migration of 1,000 of Ashton's population, and this at a time when the prospects of employment and good wages in other industries were by no means bright. This confirms the decisiveness of wage-levels in maintaining the community.

It must be emphasized also that in discussing the reaction of the settled population of Ashton miners and their families to changing economic circumstances we have to consider the factors of their having social roots of a very strong character in the community into which they have been born. In view of this, the loss of 1,000 in population is highly significant. It is beyond doubt that the temporary upsurge in recruitment in the mines in the early 1950's was greatly determined by growing difficulties of employment in other industries. Again, in the inter-war years, when the status of mining and miners went to its lowest, who can say how much of mining man-power would have been lost but for the low level of

[1] Mr. Arthur Horner at the Annual Conference of the National Union of Mineworkers. July 1954.

prosperity, albeit not so low as in mining, in other occupations, particularly for unskilled men? Clearly the industry is in a highly dangerous position if it maintains a level of wages in conditions sufficient to attract only those men dissatisfied with other industries. The National Coal Board and the nation itself can perhaps be thankful that the additional factor of the strong ties of mining families to their communities has been present to help maintain the industry's man-power at its present precarious level.

In Ashton the family is a group consisting of wife, husband, and children which subsists on a level of wages common to most of the working husbands of the town; within this level itself, as has already been shown, there exist significant differences of income. The boy brought up in any one of these families is typically destined to be a miner. Even if he or his parents have ideas of other kinds about his future, they are up against very strong factors. The much greater availability of employment in mining allows only a small percentage of school-leavers to work elsewhere without going very far afield. To take his place in the community, to share the continued friendship and co-activity of his boyhood friends, a young man cannot for long stay outside of mining in the other occupations such as builders' labourers, general labourers, etc. These jobs simply do not generally afford the prospect of the continuous level of high wages necessary to keep up to the same standards of leisure-time activity as the miners themselves. Young women in Ashton may have prejudices against marrying miners, because of the memories of 'hard times' in the pits, and the dangers of mining, but they generally marry miners, because only in this way can they assure themselves, in Ashton, of the standards they expect and know other people to expect. For all these reasons, which add up to the usual over-simplified and over-generalized view that 'people just get used to a certain way of carrying on', Ashton continues as a community of miners and their families. For the same reasons, though interviews with miners and their wives show that over 70% would not encourage their sons to be miners, the young men of Ashton are still miners!

When boys leave school at fifteen they do not always enter

mining. For a few years they will try all manner of other jobs—shop-assistants, errand-boys, drivers-mates, building labourers, etc., etc., but before they reach maturity they answer the call of the mine, apart, that is, from the limited number that can be absorbed by other skilled and semi-skilled trades in the locality, and the small number who leave Ashton. As they reach the ages of 18–21 the factors already discussed begin to take their toll. Their friends in the pit are within sight of a really good wage standard, and are already earning well above the rates paid in most other jobs for boys either apprenticed or without technical training. Also, the prospects for further education and training in mechanical and electrical engineering, and in mine-management, give mining a better outlook for the future than other jobs locally available.[1] An additional factor of apparently growing importance is the fact that from the point of view of the national service call-up, mining is a reserved occupation; many young men go to sign on at the local colliery only a month or two before their call-up is due. The following two incidents are examples of a common type.

"The training officer at a local colliery—he has the job of engaging young workers—is interviewing a 17-year-old who has applied for work.

" 'Why do you want to work here, lad?'

" 'I'm fed up with my own job. It's a dead end. I don't get enough pay. My mates work here.'

" 'Why didn't you come here when you left school?'

" 'Because my dad always told me a man who worked down a pit was a fool.' (His father is a miner.)

" 'You're sure you haven't come here just to get out of being called up?'

" ' Oh, yes, that's why I have come *now*.'

" Slightly taken aback by the boy's frankness the training officer smiles: 'You can start next Monday, lad.'

The second conversation took place in an Ashton commission agent's office.

" Two youths well-known to the clientele appeared at 2.30 p.m. 'in

---

[1] The truth of this is further illustrated by the entry of more and more young men (under twenty) coming from Leeds itself, into collieries near the large industrial city of Leeds, as well-paid jobs in other industries are becoming more and more difficult to obtain than they were in the immediate post-war period.

their muck' and everyone present knew that they had neither of them previously worked in the pit. One collier addressed one of them.

" ' Hey up, there, what's tha' doin' lookin' so bloody black?'

" ' I've started at ti' pit.'

" ' Well, tha' looks all right–sayin' tha'd never work down there and then signing on.'

" The lad replied quietly as he sat down: 'I don't care, they're not havin' me int' bloody army for two years.' "

There followed a spirited discussion among the older men present on the pros and cons of foreign policy, patriotism and conscription. In this discussion neither of the youths took part. They sat aside and listened, slightly amused at the vigour and loudness of the older men in the argument.

Neither of these youths had thought a great deal of the principles involved in the call-up, and fear of military service would certainly not have driven them into the pit. A more correct interpretation of these cases seems to be that in the context of all the influences drawing young men of Ashton into pit-work, the inconvenience and supposed waste of time involved in national service helps to force an early decision.

Our starting-point is the family as the centre of the intimate lives and relations of Ashton miners and their wives, a centre where the miners and miners' wives of tomorrow are reared and orientated towards that rather limited world which Ashton offers. It is important to ask what strength and independence this institution has for its members, how it supplements or conflicts with the other social relations encountered by its members and with what success it satisfies those fundamental needs which it has the function of serving. The latter problem is a vital one, for the family has the role of answering the emotional requirements of its members, of organizing in a mellow atmosphere the necessities and comforts of daily living for parents and children, and last of all, most important for social investigation, the family is above all other groups the primary influence in the process of 'socialization'. In addition, therefore, to a description and analysis of the relation of the family to other activities and institutions, it will be necessary to devote attention to the family's internal organization and structure.

DIVISION OF ROLES IN THE FAMILY

An earlier chapter[1] on Ashton's ecology has given examples of the detailed content of ordinary households. Normally houses are of the general working-class type, with small rooms, and in Ashton, no gardens to speak of. They are thrown together with little aesthetic taste. The arrangement of blocks and streets gives every sign of a mechanical process of providing accommodation. Standards of material comfort do vary a good deal,[2] though at the upper limit they do not approach those of the urban middle class. Nor at the lower level are they as bad as those of the typical urban slum. Nowhere does one find luxury; the qualities insisted and remarked upon are cosiness (i.e. a combination of warmth and comfort), tidiness, and above all, cleanliness, achieved through due diligence on the part of the housewife. Working people in Ashton adhere therefore to these basic standards as the main requirements of a household, which must provide a sound and comfortable place to eat and sleep for parents and children, a place where they can enjoy privacy if they ever feel in need of it, and, very important, a haven for the tired man when he returns from work; here he expects to find a meal prepared, a room clean and tidy, a seat comfortable and warm, and a wife ready to give him what he wants—in fact, the very opposite of the place he has just left, with its noise, dirt, darkness, toil, impersonalism and no little discomfort.

This ideal of the home is a simple one in appearance, but the description of its achievement in practice involves some account of the social roles of the different members of the family, and in particular of the specific division of responsibility and tasks in the maintenance of the standards required in the home.

Writers on the family in the U.S.A. have long since noted the changed function of the family and the home in the life-activities of man and wife. Industrial capitalism shattered completely the traditional state of affairs where the household was the centre and focus of the economic and social roles of its members, where it was often not only the unit of consumption but of production, and

[1] Chapter I, 'Place and People'.   [2] *Ibid.*

where it was for the individual only the point of most intensive concurrence of a widespread net of kinship ties, duties and obligations. This tendency has been most clearly discerned, and this is because it is in their experience more significant, by American writers,[1] who have had before them the example, very close in history, of the traditional pioneer family as contrasted with the typical urban family.

In Britain the transition took place much earlier and it could not be noted by historians with an ease similar to that of the modern American sociologist simply because they were not living in a period when such questions had been brought within the purview of objective study; they were rather in the province of morality.

Although this contrast of the modern working-class family with its peasant ancestor may appear irrelevant here, it is worth noting that the sociologists, by the very use of this contrast, have exposed some of the fundamental features of the modern family, features which might otherwise have been taken for granted. We are here concerned in particular with the fact that the family has lost all functions of general social significance except those of the repro- duction and a major part in the socialization of children.

THE ROLE OF HUSBAND AND WIFE IN THE FAMILY

A man's centres of activity are *outside* his home; it is outside his home that there are located the criteria of success and social acceptance. He works and plays, and makes contact with other men and women, *outside* his home. The comedian who defined 'home' as "the place where you fill the pools in on a Wednesday night" was something of a sociologist. With the exception of a small minority of men who spend a good deal of time pottering about with household improvements or are passionately interested

---

[1] For the families of miners, of course, the contrast with the peasant family is long-lived and complete. Historically the miners were the first workers to engage in that kind of work and community which determine in large measure the shape of relations in the modern urban family. The means of production were isolated from the home, the miner separated from his family for long periods, the worker divorced from any share of the means of production, working only for a wage, long before the peasant family broke up in Britain as a whole. Here again the very physical conditions and necessities of operation and working in coalmining accentuate the general features of employment in capitalist industry. Not only is the miner more intensely affected by the nature of such work today than almost any other category of workers, but this type of work, and the family it produces goes back much further in time for miners than for other workers.

in some hobby, or are very newly married, the husbands of Ashton for preference come home for a meal after finishing work and as soon as they can feel clean and rested they look for the company of their mates, i.e. their friends of the same sex.

The wife's position is very different. In a very consciously accepted division of labour, she must keep in good order the household provided for by the money handed to her each Friday by her husband. While he is at work she should complete her day's work –washing, ironing, cleaning or whatever it may be–and she must have ready for him a good meal, unless he is one of those who eats canteen meals, but this is not yet common in Ashton. The wife's ability to complete these tasks by the time of her husband's return from work is very commonly under discussion. The miner feels that he does an extremely difficult day's work; he makes it plain that he thinks it 'a poor do' if his wife cannot carry out her side of the contract. The wife is invariably found to support this view strongly. Housewives boast of their attention to the needs of their husbands, and of how they have never been late with a meal, never confronted a returning worker with a cold meal, never had to ask his help in household duties. If a miner returns from work on a wet day, and finds the washing crowded round the fire-place to dry, he will show a greater or less degree of anger according perhaps to his state of fatigue and the kind of day he has spent, but every woman knows that to present her returning husband with such a scene is not encouraging good marital relations. This task becomes an additional burden with the complication of shift-work.

This may seem petty, but it is only an example of the insistence of the efficient carrying out of each part in the division of labour in the family. On the other hand, it is not suggested that the miner-husband is a tyrant in this respect. He is likely to construct for his wife, in his sparetime, the kind of gadget which makes it unnecessary for the washing to obstruct the fire, or to help in saving for a gas cooker or an electric washing machine to help his wife. With the given resources she is expected to do the job. Another typical example of this rule is the customary practice in Ashton and other places, in refusing a meal–the method is to 'throw it to t' back o' fire' (with reference to the old-style fire-place which had a space for

reserve coal behind the fire). One man (aged 27) when presented
with 'fish and chips' from the nearby shop on returning from
work, threw them into the fire. His wife's job was to find time to
cook a proper meal for a working-man, not "a kid's supper on the
street corner", as he put it. The same man's sister and brother-in-
law (J. B.) were sharing a house with another couple. On one
occasion when J. B. returned from his work he was presented
with a good meal by the other woman in the house, who had
cooked it as a favour for his wife, anxious to go to Castletown for
the market-day. J. B. boasted afterwards that he had no complaints
about the food, but he had thrown it "straight to t' back o' t'
fire", and that when his wife arrived she was forcibly told that he
had married *her* and he was going to have his meals cooked by *her*
alone–and he stood over her while she cooked a dinner, three
hours later! Again, this man does not show by this that he is a
tyrant. He was forced later, because of deafness, to work on the
pit-top at low wages, and his wife now goes out to work at 7 a.m.
returning at 5.30 p.m. J. B. helps her in all manner of ways in the
house, and has a meal ready for her when she returns from work
each day. The strength of tradition is shown by the fact that he
still calls the performance of these duties 'helping his wife' and
that he looks a little shamefaced when it is mentioned in his
company that he makes the beds and prepares his wife's
dinner.

Young women in Ashton see their future in terms of being
married and running a household; they have no prospects of professional or other social interests and activities outside the home.
The normal state of affairs in working-class families where the
wife is a housewife and the husband a breadwinner is accentuated
in Ashton, where the main industry cannot employ female labour
and other jobs for women are scarce. The wife's confinement to the
household, together with the acceptance of the idea that the house
and the children are primarily her responsibility, emphasize the
absence of any joint activities and interests for husband and wife.
It is very unusual for an Ashton family to arrange some sort of
social event in their home and jointly entertain friends. Not only
are their houses rather unsuited to this sort of arrangement, but

the culturally established centres of leisure-time activity are located outside the home, and for the most part they are exclusively male institutions. Some husbands prefer to be outside the home when they are not working, eating or sleeping. Staying at home bores them; they prefer to have a drink or play a game of darts with their mates in the club. Television seems to be making a slight difference in this respect, but it will be recognized that as a 'family activity' this is a passive and silent one, a relation between the television screen and each individual rather than between the family members. Even those husbands who do not do a lot of 'clubbing and pubbing' do not pursue many joint activities with their wives. According to what shift they are working and according to their interests they may spend interminable hours in their allotments or on a seat in the park. Many miners are competent 'handymen' in the house and spend hours making furniture and household gadgets. Another husband will possess a motor-cycle, and his wife will complain of the time spent endlessly dismantling and reconstructing the machine. The point about all these and other activities is that in no case do they demand co-operation or encourage the growth of companionship between husband and wife. As the years go by, and any original sexual attraction fades, this rigid division between the activities of husband and wife cannot but make for an empty and uninspiring relationship. The picture of this relationship only becomes clearer when we describe the total of the traditional and practical division of duties and responsibilities in the household.

If the marriage relationship is characterized in a very general way, by a business-like division of duties and work to which the development of affection and companionship is accidental, of what does the division of labour between husband and wife consist in practice? It should be said that like so much of life in Ashton, and working-class life everywhere else, such a division is not conceived of in its long-term aspect at all, but is lived, acted out, on a day-to-day, and in this case, a week-to-week, basis. This constitutes a position where general problems and tendencies of the relationship are not considered, at least until they take the form of dominating a situation which has become intolerable.

The rhythm of domestic life is the rhythm of the working-day, the working-week, and the weekly wage-packet.

Of her husband's work at the pit, a wife in Ashton knows very little. She will hear all her life conversations between her father, brothers or husband with their friends, much of it about the pit, and yet she will rarely have a realistic picture in her mind of the work of the miner. No women are employed underground, and very few wives have had the opportunity of making a trip down the mine-shaft. In this way Ashton differs from those industrial towns where women have invaded the realms of production, where even if they do different jobs from their husbands, they experience in some degree the social relations of industry. Productive work in Ashton (and this is emphasized by the arduous, dirty and particularly 'masculine' character of pit-work) is with a few exceptions the exclusive realm of the men folk. So far as his family is concerned, he goes to work, preferably every day, and brings home the wages on Friday. The function of the home in respect of his work is to see him off in a fit state and have satisfactory conditions to which he can return. We have discussed elsewhere the quality of workmanship in the pit.[1] The miner's wife can have nothing to say on this score; her interest is in the weekly regularity of an adequate income. She is therefore interested more in seeing her husband work regularly rather than in thinking about how hard he works. One of the first requirements of a good husband in Ashton is that he should be a regular worker, and this is what wives mean when they speak of a 'good worker'. This means that the man referred to is a good worker from his family's point of view; he is rarely an absentee.

One hears constantly significant jokes about wives driving their unwilling husbands to work, but the wife's anxiety about whether her husband will 'have one off' or not is rarely openly expressed to her husband. D. N. was a 'panner' of 28 or 29 years of age, with two small children. He worked 'nights' and 'afternoons' in alternate weeks, and though for a month on end he would work regularly, with overtime at week-ends, as a prelude to Christmas or some approaching financial problem, his normal method was to

[1] Chapter II, 'The Miner at Work'.

miss one or two shifts in a fortnight. His wife recognized this, and she knew that he was much more 'absenteeism-prone' when working the night-shift. The temptation to stay away on a Monday night was exaggerated by any or both of the following factors. Firstly not being required for work till 10 p.m. on Monday, when having finished the previous week at 8–10 p.m. on Friday, makes for a long week-end at the end of which he had acquired a taste for freedom (for a man working days and nights alternately, i.e. finishing at 1.30 p.m. on Friday this tendency is naturally stronger). Secondly, and only this turns the first into a powerful influence, if the Friday saw a good wage-packet, and the miner still has some pocket-money left, he is tempted to go out on the Monday evening rather than prepare for work. D. N.'s wife like many others, could discern from all the small mannerisms and moods of her husband whether or not he intended going to work. By six or seven o'clock, she would be more or less sure of his intentions. Care must be taken not to ask him directly "Are you going to t' pit tonight then?" before his mind is fully made up–or before the time when he is prepared to admit he is not going. Perhaps at the time she normally begins to place his clothes before the fire or put a meal in the oven she will ask if she ought to do any of these things.

Of course there are variations, and the above procedure only takes place when the wife is a little worried about money, thinking about bills that must be paid. A couple who are 'flush' for the moment might joke about whether or not the husband is going to work. D. N.'s wife would confide in an outsider that she was a little worried about money and she thought her husband was almost decided to 'have one off' yet she knew that he had a thankless job, and what a sharp reply she would get if she asked him outright (certainly not tell him she thought he ought to go!). Therefore she sat apprehensively or went about her work, watching and waiting for some sign of his intentions. She also knew very well that one day off means usually at least one other in the week. The difficulty of driving oneself back to the pit after a long week-end is exacerbated by missing another day, even though the husband's money will not by now be jingling so freely. In recent years, however, the tendency to take off two shifts rather than one has been made more common and certainly more logical by the 'bonus' system introduced with the five-day week agreement of 1947.[1] A panner will say that he might as well lose £6 for two days as £4 for one.

In certain conditions the wife will even encourage her husband to 'have one off'. Many a wife will say that her husband 'goes sour' if he works too long without a break, so that he only has to say he could do with a rest for her to suggest missing a day's work. She will, of course, only do this when it will not seriously endanger her household purse, so that such a relationship is likely to exist only in the families of contract-workers, and for those of childless couples, where shift-work is depriving the young wife of her husband's company in leisure hours. Such a recently married wife knows in her heart that she will not have the prospect of 'going out' and enjoying herself with her husband for many years, and so she will often encourage him to take a shift off work. Similarly in the later phase of marriage when the children have left home or are working, a wife will reach the stage where she can safely encourage her husband's occasional absenteeism.

THE WAGE-PACKET

Apart from these exceptions in the very early and the later phases of married life, the size of the wage-packet, and its division between husband, on the one hand, and wife, family and household on the other, is the all-important basis of interrelation between family and work. The surface-worker and the ordinary day-wage men can earn a living only by working a full week.[2] Even though their work is usually more dull and uninteresting, if less arduous, than that of the contract-worker, they find it difficult to take days off work, and any tendencies to absenteeism are obviously more likely to cause friction between husband and wife than in the families of contract-workers. Many day-wage men, of course, are in their late middle-age, and the most difficult period of bringing up a family is over, so that the drive from the wife for a steady income is not so intensive. In addition such a man will during the

---

[1] Chapter II, 'The Miner at Work'.   [2] *Ibid.*

course of his years as a face-worker, have developed habits of regularity of work which have become accepted by his wife; for both these reasons, the impact of regularity of work, on marital relations in this older group of day-wage workers does not greatly differ from the position among face-workers and their families. But for those men who work at the lower-paid jobs through the early years of marriage, the period of 'getting a home together' and raising a family, it is certain that the possibility of tension and trouble between spouses is more likely than among better-paid workers. Whether or not such a division is true for working-class families in general, in an industry depending on workers from closely-knit communities, and with a not unnatural tendency to patterns of absenteeism, this direct link between work and wages on the one hand, and a threshold of trouble in the family on the other, is open to everyday observation among the miners of Ashton. Whether or not unhappiness exists in any particular marriage will depend on all manner of individual features, and many marriages will be as happy or happier than those of higher-paid workers. Nevertheless it is a question of greater or less chances of tension at different wage-levels.[1]

An Ashton wife judges her husband firstly on his ability to give her household security by means of a steady weekly income. This however does not mean that she sees the wage-packet every week and they decide jointly on its expenditure. In the majority of cases there is a standing agreement for the husband to pay his wife a constant amount for housekeeping, keeping the rest himself. The primary advantage of this system, so long as it works without hitches, is that it insulates the expenditure on family necessities from the vicissitudes of the miner's income from week to week. A contract-worker might give his wife £7 10s. each Friday: this he will undertake to do even if his wage will amount to £12 one week and £9 the next. Whatever happens he gives his wife the agreed sum, which is usually called, significantly enough, the wife's 'wage'. Of course, the sum is reduced in those cases where a

---

[1] Statistical evidence would be desirable, but clearly the evidence would still be open to question. The individual observer cannot measure marital tension, and in such a small population divorces and separations are not sufficient to use as an index with any statistical reliability.

man is kept from his work by sickness, or is forced to work at rates of pay far below his customary wages. Above this level, the wife will expect to receive her 'wage'.

Looked at from the opposite point of view, which finds just as much evidence in fact, this financial arrangement has severe disadvantages in that very often a large amount of a man's good wages is frittered away on his amusements and the wife has little say in saving. In a good stretch of work a collier or a ripper might earn regularly £13 or £14 per week, yet retains the arrangement of giving his wife £7 or £8. Should he then 'hit a bad patch' through bad conditions or through suffering bad health, he will not normally have put by any savings, and must make inroads on the 'wage' of his wife, so that the whole family and household suffer.

In the everyday (or rather every-week) lives of Ashton families there are examples where each extreme of advantage and disadvantage will dominate, and any one family will at different times feel the impact of both. However, in the give and take of 'managing' from one week and one month to another, couples adapt themselves to a satisfactory routine which guides them between the two extremes. A few typical examples will make this clear. Many men who give their wives a wage will buy their own clothes, and make themselves responsible for saving for holidays for the family (this is sometimes done by weekly deductions from the wage at the colliery office). These are the men with high wages; those at lower levels either leave themselves a bare margin of pocket-money or prefer some extremely difficult and different arrangement (see below).

But the biggest 'relief' to their household finances made by those men who pay their wife a 'wage' is the custom of helping out with unusually excessive items on the family budget, such as a piece of furniture, a carpet, a washing-machine, or Whitsuntide clothes for the children. For such purposes a woman will mention to her husband the impending expenditure and he will make sure of two or three successively good weeks so as to make the required contribution. More examples of this procedure, and its meaning in the general conduct of family's affairs, will emerge from the description of the duties and responsibilities of the wife in the monetary sphere. A small minority of husbands among contract-workers and a large minority of day-wage men and surface-workers prefer an arrangement different from the general one so far described. In these cases the wage-packet is seen by both husband and wife and the husband takes out only 'pocket money' (anything from 30s. to 60s.). According to the particular individual agreement, he will use this for his travelling expenses and eating expenses or not, and in some cases it is only 'beer money'; even his cigarettes and chewing-tobacco for work coming out of the household fund. On the last point, many contract-workers paying their wife a 'wage' leave themselves without enough for chewing-tobacco by Tuesday or Wednesday knowing that their wives will 'stand' the tobacco as an added assurance of their going to work with a good heart. Just before the time for the afternoon- or night-shift is due to begin it is a common experience in the small 'general' shops of Ashton, to see children proffering a shilling or two-or "putting it on the slate"-for "bacca for my dad for work".

This division of the wage, by whatever means, will have particular features in different families, and the examples given of 'give and take' are typical of most family arrangements. However there is even in these families, and certainly in those where money disputes are constant, an undercurrent of rivalry between the demands of the family's well-being and the demands of the husband's pleasure. One story told with great glee in Ashton is instructive, even though it is now told as something of a joke and has had the requisite embellishments:

"A young panner, P. C. married a girl from neighbouring Norwood and went to share the house of her parents. His wife's father was an old collier and before the marriage he confided in him: 'Now lad, tha' knows we used to have to pay at t' pit for us own tools, so we had summatt off t' note if we broke a shovel or a pick. Well, I've never telled Mary's mother any different, and every fortnight or so I knock a few shillings off her wage for a shovel or a pick, even though we get 'em supplied now. Now Mary knows I claim for these things so don't let on I've telled thee t' secret; anyway tha' might as well do t' same thesen' because she knows no different.' P. C. maintains that he still does this, and his workmates joke with him about his wages every week, making suggestions for deductions. He says that all his brothers-in-law carry on the practice, and one of them while still living at home once took no wages home to his mother for a month after his pony had been killed in the pit!''

TYPES OF HUSBAND-WIFE RELATIONSHIP

Among the older miners there seems a fairly equal division into two groups. One of these consists of men taking a steady and responsible attitude towards their families. From this group come the men entrusted with public position in trade unions, on the town council, and in other organizations. They are not habitual gamblers, and they are not heavy drinkers, even though they will tell, with a laughing pride, of their young days when they dashed out in groups on a Saturday night looking for a 'skinful' of ale and a good fight. Of these men the other miners will speak in the main admiringly, pointing to their efforts of self-abnegation to educate their children and keep together an attractive household, their smart dress and their tendency to drink and gamble less than most. A few men will criticize them as being parsimonious and self-important, particularly if they hold union and council positions or are among the small number of church-goers. Such comments come only from a disgruntled minority, and this minority is the group at the other extreme of responsibility to their families among the older workers. Again it should be remembered that this more irresponsible minority of miners and their wives have passed their most difficult days of managing with money—as it happens, those years were difficult not only because they had to build the household and bring up their families but because these years were years of depression and grinding hardship. In the families of this less responsible element there is a continual haggling over money, the wife accusing the husband of deceiving her as to the money he has in his pocket, the husband trying to think of ways of 'scrounging' a few shillings without telling his wife. All the worst examples found of this tendency were among those workmen who had returned to day-wage work, had not adapted themselves to new spending habits, and had not a sufficiently smooth relationship with their wives to effect a peaceful transition to the new level of life. There are a small and declining group, though they do illustrate the extreme.

The younger married men are usually working for sufficiently high wages not to have to resort to such petty disputes, and so they can be less obviously irresponsible. The following comment by A. B., a 65-year-old retired collier, is significant of the attitude of older workers to the young, who have had such an easier introduction, in terms of wages, to married life. Of course, this division between the old generation and the young is by no means the clear-cut one that he would have us believe.

" 'Men today,' he said, 'aren't as careful about their homes and families as they used to be. They're all getting easy money at t' pit, they're used to good times. If I'd earned money like some of 'em's gettin' today, I'd have had £2 pocket money and £7 or £8 in t' bank, and rest for housekeeping. But there's some of 'em today, their wives think they've been generous when they've handed over £8 or so on Friday, and they've had as much left in their own pocket. There's a lot of 'em got into that way, if they go into a pub and they have to be pullin' a roll of notes out o' their trousers pocket and peelin' one off (he gestures). If I had my way they'd all have to walk around for a month or two wi' t' dole in their pockets – and any as grumbled 'd lose that – they'd soon learn their lesson. Why, in my day, a man had to be careful for every little thing, making every shilling where he could. Me, I used to get a lot out o' gardenin'; we'd grow all us own vegetables and I used to win lots of prizes at shows. A garden was a necessity then. Nowadays men'll rent a garden, pay for it, and after a year, or two they've nought on it but weeds and thistles, and if you ask 'em why, it's "well, I haven't time, I'm working!" – (He laughed aloud) – They all ought to be fined.' "

A. B. is right in saying that the young married men of today do not as a rule give the greater part of the share in the industry's comparative prosperity to their families. For the present moment, so long as the husband can give an adequate 'wage' to his wife and still have sufficient to enjoy himself, no conflict will ensue, except a nagging anxiety and strain among the lower-paid workers. The next test of the conflicting demands of the household and the husband can only come in conditions of economic difficulty. That is, when the husband's weekly wage is not adequate to maintain living standards and to allow the husband the traditional pleasures. The conditions of Ashton family life in the depression period of the thirties are an important part of the study of the family today, since the young married couples of today were born and brought up in families living in depressed conditions. In certain ways there has persisted a tradition of behaviour in marriage from those days to the present.

Many married couples in the Ashton of the thirties seem to have adapted themselves quickly to the low level of wages, realizing that the fripperies of life must be given up.[1] In these cases husband and wife co-operated in keeping their heads above water, which was all that was possible in that period with long-term unemployment and the long years of only one or two days work a week. At this time there grew up that tradition of groups of men in the mining areas sitting or standing together for hours, with no work to do and yet no money to enable them to enjoy the resultant freedom. In interviews with middle-aged and older miners one is given a constant impression of the never-ending boredom and frustration of the unemployed years. In these conditions, not a few men periodically 'broke out', spending their 'dole' or wages before going home. In some families, not uncommon, the husband was continually guilty of this tendency and his wife was left with the responsibility of the family. All those wives who brought their families through the depression period give an impression of strength, patience, and consistency of character which asserts itself despite weariness and strain. Where the husband has not pulled his weight the wife seems to retain this strength of personality but along with it goes bitterness, or at best, a lack of sweetness in her relationship with her husband. This is associated also with an intensified relationship of affection between mother and children.

No doubt under any economic conditions there will be responsible husbands and irresponsible husbands. What the depression seems to have done is to sharply distinguish the extremes of these attitudes, bringing out clearly the tendencies of husbands along very definite customary lines. The result is a division into (a) those marriages where the husband has arrived at a position of conflict

[1] On this question, it must be realized that only the evidence of the memories of Ashton residents is available; there is no record of direct observation. The examples given are of this sort, and for that reason this section on the family in the past is kept brief.

with his family, in terms of money, (*b*) at the other extreme, those
marriages where the husband can be relied on to 'put his family
first', (*c*) the largest group, the 'compromisers' between the two
extremes, where the husband is pulled in two directions never
quite settling for one or the other. Naturally then his relationship
with his wife is a fluctuating one; she will say when asked if he
has been a good husband–"I can't grumble."

Having pointed out that the best and worst examples of marital
relations in the thirties are extreme, it will not be misleading to
give specific examples from interviews as illustrations of how the
tendencies at work were actually manifested in daily life. The following is a report of the unsolicited life-story of the widow of a
Calderford miner (A. P.) :

"I can honestly say that from 1931, except for six months at the beginning of the war, he never went a single afternoon without going in the
bookie's office and right through the years on National Assistance he
never had less than £1 pocket-money–always 18/- on cigarettes and
he never once failed to send in his football pools.

"At this date, she said, began their nine years on the dole. Financially
this was a constant struggle between A. P. and his family. In the first
months he would often collect his dole and not come home till mid-
night, depending on his fortunes at baccarat and solo-whist in 'Rosie's
Club' (later closed down as a gaming-house). Taking courage from her
desperation at trying to keep her two children clothed and fed, she
threatened to go either to the police or accompany him to the dole-office.
She did the latter, and what upset him most was the fact that he was the
only man at the office whose wife insisted on coming with him. On this
first occasion he handed her £1 and promised her 'a good clout' when
he got home–a promise which he carried out and which precipitated
a fight.

"For years J. went to the dole-office to draw the 'wage' from A. P.;
she tells stories of catching him trying to get out through the back-door
to avoid her on his way to the club. In any case by Monday each week
he was 'borrowing' two or three shillings from her. By Thursday he
was looking for any single thing in the house that he could pawn.

This anecdote is only part of a very instructive story and is chosen
to illustrate here our point about the money-antagonism between
the husband and his wife as representative of the family. Our
second example of a similar kind is from the family life in 1953 of
an Ashton miner whose present relationship with his wife is very

obviously a result of the kind of conflict described in the previous case history.

"F. G. is 63, his wife 58; he has for many years worked at good contract rates, only recently moving to 'bye-work' as a prelude to retirement. He is still very firmly attached to the habits of his well-paid days; he represents the extreme of that small number of miners who put their beer and betting before anything else. On coming home from the pit he washes and, sometimes before eating a meal, visits the 'bookie's office' round the corner. Between races he goes in the public house next door. His wife speaks to him only when it is absolutely necessary and for days on end they will show only bad temper towards each other as a result of his taking a day or two off work. In the summer of 1952 he was absent from work for three months through injury. Each Friday he would draw about 30/- tax rebate and adjourn to the bookie's office, more often than not losing every penny. He and his wife keep a number of hens, and F. G. is well-known at the pit for boasting that he has taken eggs from the chicken-run and sold them without her knowledge. His expressed views on his relationship with his wife makes it clear that he regards the marriage as a never-ending tussle for advantage, never as a co-operative effort."

An example of the marriage of an older miner at the opposite extreme is that of G. H., 57 years old, for fifteen years an official of the trade union branch, and at present contracting as a back-ripper. He lives in the new estate at Sutton.

" ' On Saturdays and Sundays I go in the best room at the Crossways. On Sunday I always take the wife and she talks to X. Y.'s wife (X. Y. is his workmate) while we talk together. Any ordinary night I'll go to the Crossways for two pints at about half-past-nine, never more, never less than two pints. Once or twice the wife and me go to the pictures in Castletown, and then I'll call in a bit later at the Crossways.' When asked if he backed horses: 'No, when I was a young man, I was very keen, and just after I got married, when we were out of work, my brother and I set up as bookmakers, but we lost everything. Yes, in those old days, I had big bets, but for years I haven't gambled. Two or three times a year I'll say to the wife "Now put your best clothes on, lass, we're going to the races. We're taking £5 and we can't expect to win a lot, but we can't lose any more than that. It's just a nice change, a day out." ' "

Here is a man, who unlike S. T. or A. P. has settled down to a very steady weekly routine which he enjoys, at the same time having a good understanding with his wife, allowing her to share many of those activities which many husbands regard as only to be shared with their 'mates'.

Still today in that part of Ashton known for its backwardness and its nearness to slum conditions, there are families with reputations among neighbours for their 'rows' which periodically break out into open fighting. All the evidence indicates, however, that this violence in marital relations has declined considerably since the years of depression. An interesting fact is that Ashton people hold the view that tendencies of irresponsibility and failure to conform with normal standards is a trait which 'runs in families', and it does seem that men and women brought up by parents who were constantly 'rowing' and fighting tend to bring this kind of behaviour into their own married life. In these cases their bickering will break out into a 'row' every few days, and the occasion is more often than not a money dispute.

In the majority of cases, i.e., those families not at the extreme of hostility between husband and wife, rows do break out periodically but they are soon made up and do not really endanger the relationship. One girl, 18 years of age, married only two months, replied thus when surprise was expressed at her 'having a row' with her husband: "Oh, that's nowt! We have a row regularly every Saturday when I ask him for my wage and he doesn't want to take me out with him." She regarded a row as a natural and regular feature of married life, not as something upsetting. Neighbours commented on what a happy couple they were. The young wife (B.) concerned in this example, gave an enlightening account of her girlhood life at home. Her father, a very quiet and morose man, used to take her mother out only on a Saturday, and they inevitably quarrelled on their arrival home. The mother knew from experience the violence of her husband's temper and would call B., one of her three daughters, from bed. The father's instructions were the direct opposite and B. speaks of many hours spent shivering at the top of the stairs in anticipation of being called to her mother's aid. Their father had found them similarly on their mother's side in other matters. When he took a day off work the mother and daughters would say that they were certainly not going to see him stay at home and live off the daughters' earnings, and so the daughters 'had one off' too.

With effects of varying strength, then, there are tendencies at work in the social life of Ashton which create a potential division between a husband and family. The claims of his spending-power are not the only factors to be considered; it is rather a matter of his participation in social relations not entered into by the rest of the family. One reflection of his participation is the division of the wage. Its importance for a discussion of the family rests primarily in the fact that the husband's duty to his family more or less ends at the handing over to his wife of her 'wage'. It will be essential to treat in some detail the whole question of the different areas of social participation of adult men and their wives. First it is proposed to complete the description of duties in respect of the wage-packet.

HOUSEHOLD FINANCES: THE WIFE'S RESPONSIBILITY

If the husband's duty to his family goes little if at all further than delivering part of his wage each Friday, here the duty and responsibilities of his partner begin. It is for him to earn the money and for her to administer it wisely. In actual fact, this means that the wife takes virtually all the responsibility for the household and the family, and the core of this responsibility is the administration of the weekly finances. Many an Ashton husband will be heard to excuse his escaping these responsibilities with such a half-jest as "I wish I could swap jobs with our lass (the wife) – it's the best job in the world, stopping at home all day." If on a rare occasion his wife or one of her friends hints at the arduousness of looking after a family, the husband will jeeringly, though good-humouredly insist that they spend hours each day 'callin'' with their friends in each other's houses.

Most contract-workers give their wives a wage of £7 or £8 weekly, and the wife of a day-wage or surface-worker will receive somewhat less than this, anything between £5 or £7. The wife of a contract-worker or other well-paid official or skilled trades-man will be able to rely on an adequate regular income regardless of the fluctutation of her husband's wage. The wife of the poorly paid worker, on the other hand, will find herself trying to manage with varying additions to the minimum which accrues from a normal week without overtime. Sometimes in fact these additions due to overtime working do not reach her. Nor is this regularity of income of the contract-worker's wife her only advantage in bringing up her family. Virtually the whole of a wife's 'wage' in Ashton is spent on food, on the immediate needs of the week, so that by Wednesday and Thursday the pennies are being counted, and most things bought in the shops are 'on tick'. No cases have been encountered where savings are made out of the wife's allowance. When any large item of expenditure crops up the husband makes a special grant from his portion of the wage. For the most part, the wife's wage is spent entirely on regular weekly items needed to keep her family fed and the household clean. The position whereby extras are paid for by the husband obviously favours the wife of the well-paid worker; she can always rely on her husband to find the required amount, provided he is working regularly and is given sufficient warning. In certain cases, if her husband is at all thrifty, he will have something saved in case of his sickness or injury; whereas little can be saved by the poorer-paid husband. The fact is, however, that even most contract-workers will admit to having no savings.

Most housewives do their daily and weekly shopping in Ashton itself; a visit to the nearby towns of Calderford and Castletown on market-days once a fortnight, and every two or three months a trip to Leeds, is the routine of most wives. For the most part, money spent during the week goes on food and it is paid for across the counter on one or two occasions, in payment of the weekly accounts which most families have at one or two grocers' shops. This paying 'on tick' is the traditional way of supporting a family. It is often extended beyond food to the buying of clothes, furniture and other household needs. The wives who can afford to buy all their requirements in cash are very proud of the fact, and to be in this position arouses some admiration. When questioned, nearly all husbands and wives say they would prefer to be able to pay cash. They dislike a large weekly commitment on furniture–a constant drain on the weekly wage. Wives who only just manage to keep going on their weekly allowance know quite well that the next Friday's wage is 'spoken for' to the last shilling, and many of them keep sums of money separately for such purposes as school dinners, rent, insurance payments, and bills at the local shops. Any Friday sees women, alone or with their small children, waiting in the vicinity of the pit gates for their husbands to give them part of their wages before they descend for the afternoon shift. A man on the night-shift will often be pressed by his wife at Friday midday to hurry to the pit for his wage so that she can do the week-end shopping ("and don't stop at the club on the way back!"). Like the wife who pays 'cash-down', the wife who is not empty-handed by Friday morning is not a little proud of the fact.

Whereas the food of the family can be bought on a day-to-day and week-by-week basis, the larger items call for some foresight and saving. We have said earlier that family life follows from one day to the next on a 'hand-to-mouth' basis. There is little thought for the morrow, little planning or anticipation of future needs except on a very short-term view. The following example illustrates the way in which typical Ashton families provide for their material well-being.

S. T. is a contract-worker of 30 years of age, with a wife and two children aged 3 and 4 years. His weekly wage varies between £8, which he draws for three shifts worked, and £15 which is the reward of a 'full week' under the best possible working conditions. He guarantees his wife £7 10s. 0d. even when he has missed a shift or two, though in the latter case he will be borrowing a shilling or two by Monday; this loan is added to his wife's allowance the following week. Of her £7 10s. 0d. his wife saves nothing; this money will be spent on food, weekly bills, and small items of clothes or cheap toys and delicacies for the children. On this 'wage'–and in this they are typical of the better-paid families– they eat very very well, though neither luxuriously nor even economically. Wives in Ashton are like those of the working-class in general; they have been brought up in the tradition of good plain food and plenty of it. Now that some wives are in a position to go a little further, they have gone in for more of the slight luxuries they have always known rather than for any radical change in eating habits. A miner's family today will have sweet cake, tinned fruit, and prepared meats much more often than they could afford to buy before the wars (though in the lower-paid families fresh meat, bread and potatoes are still the everyday fare with little variation).

One notices very easily in Ashton that fresh fruit and vegetables are often neglected when preserved foods, much more expensive, are bought. This is due, of course, like the persistence of traditional leisure pursuits to the fact that the advance of the mines has been largely confined to the wage-packet, and the increased amount is utilized according to a way of life founded on the old level of wages.

We return to our example of S. T. and his family. Food is here bought and paid for each Friday at two shops, the local Co-operative store and a small 'general' shop only four doors away from the house. Two visits to the Co-operative stores complete the purchase of the main groceries of the week, and through the week the children are sent out to the nearby shop for loaves of bread, a few biscuits, cigarettes and tobacco, or sweets for themselves. On this basis the meals of the family do not vary from one week to another, and it is just this regularity which is the wife's responsibility. She does not consult her husband about the particular difficulties she encounters, or the worries about meeting a certain weekly account; all she can do is eventually tell him that she 'cannot manage' on the amount he gives her. Again this illustrates the ending of the husband's responsibility once he hands over the weekly allowance. Over and above this maintenance of the regular weekly standard, however, it is the husband who is called on to meet the family's needs. In June, perhaps, he and his wife will discuss the fact that they have provisionally booked a holiday, and he will intimate that for two months or so he will work regularly for the necessary money. That saving for holidays is usually left late, and for the most part very late, is shown by the inevitable mammoth production figures and minimum absenteeism rates for every 'bull-week', i.e. the week's work on which the pre-holiday wage-packet will be paid.

This 'short-term' saving for holidays typifies all saving. It has already been pointed out that the initiative in saving is left to the husband, since he gives his wife only sufficient for the weekly routine.[1] The husband in our example, S. T., is like most others in that he possesses no 'general' savings. (In 1953 he was diagnosed as having pneumoconiosis and ordered to cease work. For many weeks he did not receive full compensation, and in this period he had no savings whatever to which he could turn.) Besides her weekly 'slate' accounts his wife has a quarterly account with the Co-operative, by means of which she buys large items such as, in 1953, a kitchen cabinet (£14), a fireside chair (£9) and a complete set of winter clothes for the children. Two or three weeks before the quarter day she tells her husband that they will owe say £12 or £20. In the period before payment the husband then works all the hours he possibly can, each Friday giving his wife all but a modicum of 'spending money'. After this effort he will probably miss off a couple of shifts when his turn for nights comes round.

These quarterly bills are not the only regularly occurring occasions for short-term saving. At Whitsuntide it is the custom for children to have a new set of clothes. Many families, such as the one we use as example, make each Christmas the occasion for a similar expenditure. It is, of course, possible to buy these clothes on the quarterly account, but where possible they are bought in this family for cash earned in a week or two of special effort. Many families patronize travelling salesmen, or 'packie-men'. These travellers are the representatives of a large hire-purchase clothing stores in Leeds.[2] Mining villages and small towns like Ashton have the reputation of being very fertile sources of earning for the packie-men in the post-war period. The salesmen come round the streets by car or van, calling from door-to-door to collect the instalments on goods bought previously or to persuade customers about to purchase all manner of garments, mostly children's clothes, which he is carrying about with him. The packie-man flourishes in Ashton because he is asking housewives to buy in a way in which they are accustomed and which they find more

[1] We are referring to typical families. In that minority of families with steady and very responsible husbands, saving is often on a joint basis.
[2] These stores have an ordinary retail and hire-purchase business in Leeds itself. The selling from door-to-door is carried out by specially employed travellers.

convenient than cash purchase under the conditions of a weekly wage. (In S. T.'s family the wife had made only slight use of the packie-man, since she knew she had a husband whose job made it possible for him to give her a considerable extra allowance when necessary. She did not therefore have to place any further commitments on her weekly 'wage'.) It is naturally among the lowest-paid families that the 'packie-men' find business easiest to canvass.

In summary, the family's weekly wage is strictly divided into one part for the wife, with which she must maintain the household, and another part for the husband, to spend as he will. The husband would lose the respect of his family and of outsiders if he refused his wife's requests to contribute occasional sums for responsible purposes over and above the wife's weekly 'wage'. This division of the wage between man and wife, and their duties and liberties in respect of the allotted shares, are aspects of the whole system of division between the accepted social roles of the sexes in Ashton.

THE SOCIAL POSITION OF THE WIFE IN ASHTON

Unlike her husband the wife in Ashton does not spend any money on herself or her amusement without detailed consultation with, and approval from, her husband. Some wives will bravely buy themselves clothes or ornaments and then present them for their husband's approval, but only when they know that 'no trouble' will ensue, or where their relationship has come to the point where she goes out of her way to assert herself despite the consequences. Normally all items outside the weekly household budget are questioned by the husband and accounted for by the wife. If a wife wishes to spend money on new clothes she tries to persuade her husband to buy them, which is usually successful, or she puts smaller items on her hire-purchase account and does her best to pay them off. Perhaps at a later date she may wish to ask for an increase in her weekly housekeeping allowance. At such a time she knows her husband would be likely to delay or refuse the increase on the ground that she was frittering money away on herself.

Nor in her leisure-time does the wife spend freely like her husband. Many wives, if they can arrange for the children to be looked after, or take them along, visit the cinema once, twice or even three times a week, but this is the limit of their routine spending on leisure. On odd occasions they may go on outings or a 'trip' to the seaside organized by a local club or public house.

Her lack of any personal means of spending is not the only restriction of a wife's participation in social relations wider than the family. Indeed, in cases where wives are working, the wife's earnings are invariably found to be regarded as part of the housekeeping money; thus, the economic semi-independence put within the reach of such working wives is not grasped. In the first place the traditional division of labour in the home decrees the household tasks of looking after the home and caring for the children are exclusively women's work. In most cases this idea that the wife must regard home and children as her primary responsibility bars her from outside employment at least in the earliest years of raising a family. Many a girl under 20 will say "I'll be glad when I get married, then I can stop going out to work." It is not, of course, only the social relations of work from which the mother is excluded. Leisure facilities for adults in Ashton cater very definitely for males, so that even if the wife escapes her household duties she has difficulty in finding opportunities for leisure on the same scale as her husband.

Add to these very concrete restrictions on the wife's social life their reflection in the ideology and morality of the people of Ashton. The ideas of what a woman should and should not do persist very strongly, and, of course, they are values accepted by the women themselves. No 'self-respecting' young woman will go into a public house unaccompanied by her husband or if unmarried by her fiancé. Ashton is still sufficiently small for any woman guilty of such a misdemeanour to be made to feel the weight of public opinion. Older men and women claim that great inroads have been made recently on the old conventions. At one time, they will say, women were never seen smoking or going into a public house, and no man would think of marrying a girl whom he knew to have had other lovers. "Nowadays you see young lasses smoking on buses and sitting boozing in the pubs, and the men don't seem to care how many others a girl's been with before they marry her"–so said an Ashton widow of sixty-seven. Her picture may be correct, but the impartial observer with no such impressions from the past notices above all the exclusion of women from these activities in Ashton which are outside of work and the family; these are intended for and principally used by men. Any woman who transgresses her confinement to the home and family is 'talked about' and publicly disapproved. Again the factor of size in allowing for closeness of community seems particularly important. Only two miles away in Castletown, a place of 43,000 people, more diversified occupationally than Ashton, and with marked social stratification, the presence of unaccompanied women and girls in public houses is commonplace.

Restricted to the home as they are by reason of this variety of factors, wives do not actively resent it. When pressed they will acknowledge jealousy of their husband's freedom, but many of them say that they find satisfaction in the care of their children. (Indeed the confinement of the wife to the internal affairs of the family makes her much closer to her children than the father, a point which recurs with some importance below.)

In certain conditions, in larger towns it would appear that wives are at an even greater disadvantage in this respect than the women of Ashton. The latter are for the most part living within easy reach of kinsfolk and old childhood friends of their own generation. This gives at least greater scope for social contacts from day to day than the urban middle-class wife in a city suburb, or the working-class wife on a new estate in the larger towns. These contacts with neighbours and relatives form the foremost part of the Ashton mother's extra-familial life. Every day a woman will receive a visit from one or other of the neighbours, or she will make such a visit herself. The usual practice, in fact, is for a group of three or four to gather over cups of tea and 'have five' (i.e. five minutes, which can be stretched indefinitely). At these gatherings there is endless gossip about other neighbours, about their own husbands and children, about the past. Just as the men in the clubs talk mainly about their work, and secondly about sport, and NEVER about their homes and families, so do their wives talk first of all about *their* work, i.e. the home and family, and secondly within the range of things with which they are all immediately familiar. In the evenings when the men of the family talk about the pit, and the wife or daughter sits weariedly knitting or half-listening, one knows that the conversation is more or less meaningless to her; only once was one of them heard to say (to her father and her husband) "I wish you'd talk about something I could talk about, instead of the bloody pit all the while. I'd just like to have an hour or two down that pit, just to see what you do down there." Her last remark was ridiculed, her husband saying "I know what'd bloody happen to a woman if she showed her face down there!" but he refused to elaborate for her.

The strict sexual division of labour in Ashton emphasized by the nature of mining, has extensions in the range of social contacts and the ideas available to the two sexes. Just as most women see further than the home and the tasks of its maintenance on only one or two occasions each week, the husband is very often a comparative stranger to his home. One wife of thirty said of her husband "The only time I see anything of him is in bed." Her neighbour described her husband as always being "at work, at the club, or in bed". Other men spend all their non-working hours in their garden, or, a very small number silently pottering about at home.

Besides her neighbours the typical housewife in Ashton sees a good deal of her kinsfolk. Among the older-established families the visiting of relatives is a well-developed institution. At weekends, visits will be exchanged and many wives see one or another of their kinsfolk every day. According to the relations that have developed in particular families, daughters who marry will either try to live as near as possible to their mothers or married sisters, or in some cases do their best to live at the other end of the town–"otherwise we'd never be shut of relations". A woman who takes the latter step will naturally fall back more and more on her neighbours. For a housewife living near to her relations, these form part of the circle of daily acquaintances with whom she gossips, but they will generally share and co-operate in other activities. Again we take the family of S. T. and his wife Jean, typical of families with nearby kinsfolk.

"This family lives only four doors from the household of the wife's mother and retired father. With this father and mother live another daughter (Mary aged 21), her 27-year-old husband and two very small children. They are living there until they can find a house of their own in which to live. Thus, the sister, Mary, makes at least three visits a day with one or both children. Her father visits S. T.'s household at least once daily, to borrow something, or perhaps simply to keep the child occupied. Jean makes the bed each night for her mother who is a cripple. Her two children, Stephen and Jona, obviously regard Mary and their grandmother as very close to them. Anything cooked for a meal by Jean and not eaten is taken to the mother's house to be finished. Jean says, 'We never have any letters or have to write to anybody, all our people are here.' On a Sunday her husband Stanley is always two hours late for his dinner, giving the excuse that he has been visiting his mother and sister at the other end of the town. Jean and Mary will either go to the shops together and leave their mother in charge of the children, or take turns to look after them while the other goes out. Both Jean and Mary joke about having four children instead of two. Their brother Ralph has gone to live with his wife some five miles distant in Burley, but each week-end the two of them come home and alternatively sleep in the two households. On Saturday night two or three of these couples will go out together."

Here is a family which in fact extends much farther than the biological unit of parents and children. It represents the extreme of cohesion of an extended family in Ashton, but most Ashton families have some degree of contact and shared activity with other kinsfolk. In this, as in the fact that she is close to old acquaintances of her youth and long-established neighbours the Ashton wife is less isolated, and less forced in upon her home, than many urban wives.

These customs of visiting and 'callin'' with kinsfolk and neighbours are the limits of social contact for Ashton wives; as we have seen, the organizations and institutions of the town, with whatever scope for initiative, leadership and success they provide, are based (a) on the colliery, and (b) on the leisure-time requirements of the Ashton miners. A woman fulfils herself in keeping her home clean and tidy, her family healthy and well fed. She resigns herself as a rule to the home, though there are a few wives who manage to 'keep up' with their husbands' leisure-time activities. Such women are usually very young and as yet childless, or else they have left behind

the difficult years of rearing children. They are definitely exceptional. As soon as a girl marries, unless she goes to live well away
from her previous residence, she is brought into the local circle of
married women. Like them she soon ceases to spend a great deal
of her time on clothes and her appearance; such an obvious concern with her own person would belie her new-found duty of
looking after the home. Like her equals, she learns to go to one or
another house in the middle of the morning or afternoon, light a
cigarette and chatter over a cup of tea. Like them she regards as
her first duty the provision of comfort and food for her husband,
returning from work. The fact that many girls are pregnant at
marriage[1] is a strong factor in pressing them quickly into this
restrictive mould.

It is clear that the activities of married women in Ashton are of
a very limited range, and do not allow for much expression of
qualities of intellect or personality. Certainly women have almost
the sole responsibility, within the family, for children, and in this
way they must of necessity be very influential, but the long-
established method of bringing up children and the absence of any
educational facilities for women which would enable them to do
a good job in this sphere, are barriers to any real advance. In the
mother's hands are the discipline and early personality development of the children; education is entirely the job of the school.
Beyond this task of rearing children[2] there is almost no scope for
the development of other qualities in Ashton women. There are
two women town councillors (of a total of twelve)[3] and a very few

[1] It was found to be impossible to obtain the statistics for children born in Ashton in the
first nine months of marriage, but the field-workers' own information suggests that at
least a considerable minority of first-born children are conceived before marriage.

[2] See end of this chapter 'Children in the Ashton Family'.

[3] It might be objected that one in six on the Council is a high proportion of women and
thus gives the lie to our emphasis on the restricted activities of women. There are special
circumstances to account for this proportion, the figure can not be regarded as a purely
chance occurrence in so small a population. The Urban District Council is a monopoly of
the local Labour Party, and the composition of the Council reflects the internal organization of that Party. In the past few years in Ashton there has been developed a strong
Women's Section of the Party, which in numbers of active members and regularity of
activity surpasses the male section. Although its activities apart from the occasional lecture,
are non-political, and concerned with 'feminine activities' like knitting, sewing, jumble
sales, trips, etc., the financial and numerical importance of the women's section in a local
Labour Party whose most flourishing activities are in any case social ones, has brought to the
fore in local Labour Party organization the leaders of the Women's Section, and they take
their part in the Council.

women serve regularly on local committees. Such activity and even such interest in politics etc., are outside the scope of most women. When a woman does express any interest in politics or other general topic she speaks rather apologetically, and can be prepared for her husband to tell her not to interrupt intelligent conversation –"What the hell do you know about it?" In their reading house-wives depend almost entirely on 'romances' either from the weekly women's magazines or from paper-backed novelettes.[1]

PERSONALITY AND THE ROLE OF THE WIFE: AN EXAMPLE

There is among housewives a good range of personality types at the age of marriage, but it is not difficult to notice every day how they gradually adapt themselves to the daily and weekly round of household duties and the behaviour thought correct for a wife. The two sisters, Jean and Mary, noted above, have very different types of intellect and personality. Jean never reads; she finds difficulty in following any but the simplest arguments; she has given up the idea of learning anything but is resigned to doing what she now does every day for the rest of her life. Her sex-life is unsatisfactory, but she can see no way forward, almost regarding it as inevitable; her husband tells her she is 'cold' so she assumes that to be the explanation. Her sister Mary (5 years younger) on the other hand quickly grasps new ideas, reads almost anything she can lay her hands on, discusses general questions with spirit and some knowledge, and occasionally reflects on the restrictions of her daily life.

Jean makes up for her dullness in intelligence with a talent for entertainment and 'letting herself go' in company. Mary com-bines her thoughtfulness with a bright and charming personality.

Now these two women are a good illustration of the destiny of women in this male-dominated community, for despite their varying potentialities they are both being moulded to the same shape. This statement is true physically as well as in terms of personality. One notes in Ashton, as in other mining areas, the rapid decline in physical beauty among young married women.

[1] Many of these books are bought outside Ashton on market-stalls to which they are later resold.

By the late twenties they are often flat-chested, all colour and fresh-
ness have left their faces, and they seem to be hardly concerned at
all with their physical attractiveness. Of these two sisters, Jean has
already reached this stage, and Mary, after having had two children
in quick succession before the age of twenty-one, is resigned to
going little farther than the shops and the neighbours' houses, so
she makes no effort to look attractive. One contributory factor to
Jean's physical decline is a succession of miscarriages. These she has
accomplished personally and deliberately, like many another
Ashton wife, by means of drastic dosing with ordinary laxatives.
Such miscarriages and post-natal troubles among working-class
wives who have not yet been educated to the need for constant
medical check-ups and attention, are no doubt largely responsible
for the obvious physical decline of married women.[1] To them
must be added the fact that, with birth control clinics very inadequately provided and publicized, young women are thrown
very young into a position of total responsibility in running
a home and family, and this is a considerable drain on their
vitality.[2]

When Mary and Jean were told repeatedly of the ease of birth
control under modern clinical supervision, they reacted differently
and in a manner illustrative of many general features, including the
one with which we have been mainly concerned, i.e. the conditioning of women in Ashton to a narrow and traditional range of
activity, ideas and personality. Jean was at first unreceptive to the
idea of birth-control appliances. She had a feeling that there was
something unnatural about it; she would feel awkward and embarrassed about going to the clinic. Secondly, and very significantly, she thought her husband would not favour the idea; once,
she said, she had brought home some rubber sheaths sold to her by
the chemist and her husband had thrown them on the fire, saying
that they took all the enjoyment out of sex, and were no safer than
withdrawal. She thought Stanley would ridicule the appearance
of the appliances which she was shown. She had no bathroom, and

[1] For a pre-1939 factual survey of this question see *Working-class Wives*, Pelican Books, 1938.
[2] In the whole of the area covered by the West Riding Authorities, there are only four
birth control clinics run by voluntary bodies, but grant-aided. The nearest to Ashton is
seven miles away.

she would find it very awkward to use the equipment correctly. With these objections she resisted the suggestions for a month or two. Her sister, Mary, however, saw clearly the advantages for family planning, and could see that with a little care and common-sense there need be no feeling of unnaturalness or difficulty of the kind imagined by Jean. Her husband might not take easily to the idea but she would reason it out with him. She agreed to give the matter a little thought, especially about visiting the clinic. She soon decided that she wanted to go and after a few weeks persuaded Jean. The interesting conclusion of the story is that neither of them actually went to the clinic. They never broke out of the daily routine sufficiently to fix appointments, and make the journey to nearby Bousfield. They were unused to telephones and keeping appointments and discussing intimacies with strangers. They hardly knew how to behave and explain themselves in front of the other housewives they know. The only solution was for an outsider already experienced in the clinic to make the appointment for them and take them to the clinic. Mary, although she was 'potentially' more advanced than her sister, was restricted sufficiently by her social conditions to behave no differently.

In a similar way Mary was easily persuaded of the need to exert her rights in the National Health Service but she took no more concrete steps than did her sister. "People will say I'm always running to the doctor."

Mary still has not resigned herself to the life of an Ashton housewife. She would like to go out to work, and speaks rather nostalgically of the fun and friends she enjoyed before marriage. On one occasion she expressed her wish to return to work in front of her mother, who replied, "Now, Mary, you're not going. You know how often Bill (Mary's husband) laiks a shift now. God knows how little work he'd do if you were bringing money in." Mary is unfortunate in some of the habits of her husband, but it is clear that in her particular case she is kept from outside work by the value placed on the weekly wage in relation to pit-work and living-standards.

Another example of a young woman succumbing to the limitations of married life is P. M., now aged 23. Before marrying in

1948, she was a girl of great vitality and described herself as having been 'mad on dancing'. After marriage she insisted that if her husband could go out alone for a drink or to the cinema, she too would pursue her favourite pleasure, and occasionally she would take herself off to a dance. She has not been to a dance between 1950 and 1953 because her husband was very firm, and forcibly insisted that she gave up dancing, and she says that nowadays she has stopped talking about dancing. She goes to the local cinema once a week, and has given up what used to be a regular visit weekly to Castletown with her husband now that their liking for each other has reached a very low ebb. She has actually started knitting recently, and does this, in between gossiping, on most evenings.

In Ashton, a woman who marries will sooner or later, by reason of the duties and responsibilities and under the pressure of prevailing ideas in the community, conform to a very definite type, so that among women who have borne one child or more little variation is to be found. One does not encounter women with ideas about general questions, or interested in cultural activities or the running of organizations. If a woman dresses well and makes herself attractive or runs her home at a standard far above the normal, she then receives only jealous admiration and is made to feel that she is not conforming to the accepted type of Ashton wife. One young woman, the widow of a worker killed in the local colliery, has since her husband's death worked very hard. She is now in a position to talk about sending her 8-year-old son to a 'good' school, and is known to have bought an electric washing-machine and a little 'contemporary' furniture. It is generally said of her in discussions of household standards "Oh, Mrs. X. has pots of money. No wonder she can buy these things" –and she is thereby placed beyond the limits of comparison.

THE GROUP OF ADULT MALES

It is not relevant in connexion with a study of family life to describe in detail the activities of men. The wife, after all, is essentially confined to the home and family, so that her activities are always directly relevant to the functioning of family life. The significant aspect of the husband's participation in social life is its *isolation* from his family. (The spheres of participation are described in the previous chapters on 'The Miner at Work', 'Trade Unionism' and 'Leisure'.) This isolation is reinforced by the *exclusively* male character of all these spheres of culture. The occasions when women are admitted to these departments of life, e.g. the opening of the club concerts on Saturdays and Sundays, only serve by their exceptional nature to emphasize the social division between the sexes. Men grow into a status, together with their age-mates, which makes them eligible for participation in these institutions and activities. At the same time they grow into a set of attitudes and ideas which very consciously exclude women from the activities and permitted liberties of the male group, which can be said to constitute a type of secret society.

In general one can say that there are two basic conditions for the divergence of development between the sexes. First there is the nature of work in those industrial areas where facilities for employment have always been almost exclusively male, and secondly there is the traditional role of the housewife and mother in the industrial family in this country, her confinement to the household. We have seen that in Ashton, and presumably in other mining villages, there are factors both strengthening and mitigating these two basic conditions. Mining is the extreme of exclusively male industries, and thus serves to emphasize the sex-division; on the other hand, the closeness of the community in Ashton gives each housewife some scope for a broader daily range of experiences than her counterpart in some other industrial areas.

This 'secret society' of adult males is important for a study of the family; its attraction for a man is a challenge to the growth of a full relationship with his wife, and it also cuts down the amount of time and interest which he devotes to his family. There is no doubt that the conflict between married couples springs from the antagonism between the interests of (*a*) the family, and (*b*) the husband's participation in the activities of the group of adult males. We spoke previously of the financial aspects of the husband's pleasure in relation to the welfare of his family; those pleasures are located in the activities he shares with other men. He spends his money with them in a certain institutionalized way because in this way he maintains his status and prestige in the group. That section of the group of males with which he shares his activities will often be one within which he has grown up. When arrangements are made in the group to visit a certain sporting event, or when the group goes on a drinking spree, he does not like to be left out; he likes to be thought that he is 'still one of the lads'. In the man's life the phases of his participation in this group of males are very significant for the family of which he is a member. Before describing these phases and their effects on the family it will be useful to have in mind some of the marks of the division between the sexes.

The group of men is an exclusive group

We saw that in the general meeting of the Ashton Labour Party, although women were present and even in a majority (because of the exceptional and novel development of a women's section), the men dominated the discussion and regarded the women's contributions as more or less valueless. This example points in a similar direction to the family conversations reported in this chapter; affairs not restricted to the domestic group and its welfare are regarded as fit only for men to discuss authoritatively. A woman who arrogated to herself a place in this male-dominated sphere would be looked at askance. When miners think of, say, a schoolmistress, they might make suggestive remarks about her sexual attractiveness and appearance, but they cannot conceive of her as a possible partner or companion. She is a different type of being from the women they would consider for a girl-friend or a wife. This is not just a matter of different levels of culture and education; miners evaluate a *man* according to whether or not he is 'a good sport' and can take a drink with them, and he can soon join their group despite being a non-miner.

Expulsion from the realm of ideas thought fit only for grown men is the reflection of virtual exclusion from the actions of Ashton men. The pit, the club, the pub, the bookie's office, the trade union, all the places where men do talk together, are closed to women even if this is not rigidly laid down and enforced as in the clubs. There is no rule against women sitting in the bookie's office all afternoon just as do some of the men, but women never do this; in the first place both men and women would be likely to ask them why they were not getting on with their work. But this is not the only objection; there is a *feeling* that it is not proper for women to frequent certain places.

"Mr. D. is 65; her husband is confined to his bed with tuberculosis. Now for years he has spent 1/- every Sunday backing his club pool number with the local bookmaker. He asked his wife one Sunday morning to go and place the bet for him. This she did and went to join the queue in the bookmaker's office. Soon she objected as a man pushed in front of her in the queue. Her objection was brushed aside: 'What are you complaining about, it isn't your place to be here, you ought to be at home.'

"At another bookmaker's establishment in the town, women place their bets or buy tickets only at the side of the counter, standing in the doorway. They cannot get inside the office or even get a view of who is inside, without pushing aside a screen. Women who stand talking behind this screen receive from the clerks or from anyone who sees them all manner of suggestions that their nearness is dangerous. One woman asked the clerk for winnings from the previous week. 'Where's your ticket?' he demanded. Her answer 'I didn't have it', was received with laughs from the men inside, and one of them called, 'If tha' comes in here, tha'll bloody well have it!' "

Normally no man would make such a joke in the company of a woman, much less direct it at her personally, unless they knew each other very well, and such behaviour has become customary. The reason for a woman at the bookie's office having to suffer such remarks, which she will usually take with a laugh, is a feeling among the men that she has invaded a preserve of the menfolk, and she must, therefore, be prepared to take the consequences. In one Ashton public house on a week-day a young woman was very close to a man who used one of the sexual swear-words. He was warned to be careful—"Ladies present, ladies present!" The man apologized, only to repeat the offence some ten minutes later. This time the woman's companion threatened the offender with force and received the retort: "I'm sorry, old lad, but she must expect to hear what comes out if she will come into the place at all!" This did not exonerate him, however, and he was compelled to leave. It is a rule that women keep out of those places reserved for men, but it is rightly just as rigid a rule that sexual swear-words may not be used in the presence of women.

Indeed the whole morality of swearing is of some significance, for it goes along with other divisions in thought and action. It is very noticeable that most miners swear a great deal—more or less every sentence—when they are working, when they are in the pit, and it is the most violent, i.e. the sexual swear-words which are used. Just as the miner does not go into a public house for a drink in his pit-muck he does not indulge in this violent swearing 'on top', i.e. out of the pit. When he goes in the public house he dresses smartly and respectably, and in addition he speaks respectably. In the public bar of a public house one will hear swear-words used, but not at all to the extent used at work, and the loudest and most heated and vigorous arguments will be held without swearing. Strangers come into the category of women and children so far as swearing is concerned; their presence normally forbids it. (Our remarks in the paragraph refer only to the sexual swear-words.)

One housewife described the rows of her next-door neighbours, and trying to show just how serious the situation was, she concluded "And they even use *pit-talk* when they get really mad with each other; it's terrible!" By pit-talk she meant 'filthy talk' she explained, and, pressed further, she meant in fact the sexual swear-words. The pit, of course, is the primary, principal, and most rigidly exclusive domain of the grown men, and their wives and families are excluded. In other places of male exclusiveness it is permissible and not uncommon to use the sexual swear-words but in the pit it is the *extreme* of this liberty of speech for men. No doubt pit-work provides factors of danger and nervous tension which give added incentive to swearing, beyond those situations where swearing is simply permitted by the absence of women and minors. A first glance at swearing habits suggests therefore that the group of mature men is a closed group which once entered entitles the member to certain definite privileges, among them obscenity, in given situations.

Once this distinction of behaviour in the matter of swearing had been noted, observations were made of swearing habits at a meeting-place of the two different types of situation. On Monday mornings, when perhaps a hundred men are waiting at the pit-top ready to go down to the first shift of the week, one hears in the buzz of conversation only the 'normal' amount of swearing, i.e. normal, say, in the public bar on a Saturday night with no women present. And yet already as the cage is descending one notices the growing tendency for every adjective and pronoun to become a violent swear-word, and the shift is only an hour or two old before this tendency gains dominance. Again at the end of the shift there is a change, but this time it is a more abrupt one; there is no gradual cessation of swearing towards the end of the shift. Swearing continues right until the men reach the surface and disperse to their buses, bicycles, and the canteen.

The use of the sexual swear-words in the pit is only part of a general attitude of toughness and near-callousness in conversation. Wherever they are, miners will put forward their view in an argument with great force and vehemence in a manner suggesting unfriendliness and bad temper which is certainly not present. In the pit this characteristic had added emphasis and receives the support of swear-words. Men in exclusively male company talk very lewdly about sex in general and about particular women without any consideration of feeling for the personality of such women. Men will say, pretending to joke, "A woman's only good for two things—looking after the house and lying on the bed." One collier—although his workmates showed some disgust amid their laughter—said, "When a woman's in her teens and her twenties she's worth having—she's just what a man wants. After that she's finished."

"After she's thirty, what is she?"

"She's just an old cow!"

These are but two examples of the typically loose and unrestrained conversations about sex which take place in the pit. Such conversations about women and sex are not common, but when they occur this is the general tenor of the remarks made. On one occasion when the colliers on a certain face were 'waiting for the tubs' and sitting on the coalface, a 45-year-old man in a very matter-of-fact way told to the younger men present the story of his first sexual experience. "We'd been talking and playing around with these two lasses and one of them had a long way to get home –I was only 17–and she wanted me to walk her home. I told her straight I wasn't going unless there was something at the end of it, and she said, 'Come on, then', so off we went and got down at t' side of a wall. Well I din't make a right good job of it but I had a bloody good try." His account is abridged; it was rather more descriptive and the swear-words have been omitted. The point is, one never hears such a conversation in the other places where men gather; this collier felt free to speak in this way in the pit, to his mates. Only a couple of very close friends, when the way had been well prepared, might have said the same thing, and then it would have been in a different way.

Males share exclusively situations other than the pit, and these situations occupy a position midway between the pit and the family or the other situations with females present. In these intermediate situations such a conversation would take place in somewhat milder form. Here again it is interesting that strangers take on something of the character of women and children. One example of this was the reactions of a number of miners to the film *Niagara* (Marilyn Monroe). In the bookie's office or at the pit they made jokes about the suggestiveness of certain shots of Miss Monroe, about her possible effect on certain persons present, and about her nickname, 'The Body'. Indeed any man seemed to gain something in stature and recognition if he could contribute some lewd remark to the conversation. On the other hand, in private conversation with a stranger the same men would suggest that the film was at the best rather silly, and at the worst on the verge of disgusting. Finally, the men's comments in the presence of women were entirely different. In a group of married couples who all knew each other well, the women said that they thought Miss Monroe silly and her characteristics overdone; the men said that they liked the thought of a night in bed with her. The more forward of the women soon showed up their husbands by coming back with some remark as "You wouldn't be so much bloody good to her any-way!" and the man would feel awkward. The following example

of young men and women not married and not of so long
acquaintance shows significant differences :

"A group of six, three young men and their girl-friends, engaged in
the following conversation after seeing the same film, while waiting for
a bus :
"Two of the girls did most of the talking, mercilessly 'kidding' the
men :
"I asked Jack for a light twice, but he never bloody well heard me !'
"She then selected for a re-description a scene in which Miss Monroe
is shown in a very suggestive position, apparently covered only in the
thinnest of white sheets drawn tightly across her. The girl delighted in
embarrassing the men by quoting this, and then :
" 'They showed you more of that Jean Peters than they did of Marilyn,
didn't they, Jack, love ?'
"Jack tried to blunder the way out of his blushes by an aside to Joe :
" 'They didn't show you Marilyn's legs very much did they ?' but
they both refused to make any comment on Miss Monroe. The second
girl took up the running.
" 'Did you hear them all, the men, when they showed you her the
first time ; everyone drew in his breath.'
"Until the conversation changed the girls remained tirelessly on the
offensive, taking advantage of their boy-friends' embarrassment under
the eyes of the rest of the queue."

The girls in this example have taken note of the way in which
young men normally talk about sex. They knew enough of the
usual conversation of men to touch on just the type of incident
which was familiar in the youth's ordinary conversation, and yet
the young men were helpless. The only 'context of situation' in
which they ever spoke freely of such matters was exclusively male,
with all the freedom that that allowed. With the women present
and even broaching the situation they could say nothing. On the
following day the same youngsters swore and spoke freely with
their mates on the same subject.

Swearing in conversation between the sexes and even the telling
of dirty jokes, so long as the participants know each other fairly
well, is permissible, and in many cases common. Women among
themselves make a similarly free use of all except sexual swear-
words, and there is little compunction about using the milder
swear-words in front of the children. It is at the sexual swear-words

that the line is drawn. A miner when with his wife is as likely as not to strike anyone using really bad language. Any attempt to discuss sexual problems seriously and frankly with a mixed group was found to be soon acceptable to women, who welcomed the opportunity, but the men were embarrassed and almost silent. The only context again in which they discussed sex was in a circle of jesting males. We shall return to this in a brief discussion on sexual relations in marriage.

A certain amount of attention has been given to swearing and the intrusion of sex into conversation simply because these phenomena can be clearly observed, and this serves to mark off easily those very definite realms in Ashton's social life which gain some new quality because of their exclusively male nature. It is not by any means suggested that this almost ritual division is confined to Ashton or to miners. All the outward signs are that it is typical of classes in our society. However, at higher levels in the social scale the wife is more easily relieved of the worst features of this division, those which most depress the status of women in Ashton.

From the sociological point of view, the most significant feature of those areas of social life where the men behave differently simply because they are alone, is that in every case those allowed to participate and whose presence does not restrict in any way must have definite qualifications. Anyone not possessing these qualifications is excluded, and such a person's presence immediately affects the conduct of the members. It is for this reason that women, children, and strangers are treated as though they were similar to each other. This final example on swearing shows a situation where the principal is torn between two ways of acting. P. Q., a collier of 57 years of age, described his experience in the following way :

"I'm like any other miner; I can swear as well as anybody, and, of course, my son can as well—after all he's 27 and he's working at the face. But we've never sworn in front of each other. In fact I don't think he's ever heard me swear. But one day I was sitting waiting to go out of the pit and a group of colliers came and sat nearby and he was one of them. They started talking and they swore just like any other lot. My lad didn't know I was there and so he swore as merrily as anybody else.

Well, I've never felt so awkward in my life before. I could feel myself blushing and managed to creep away without him seeing me. I'm glad I did because we'd both have felt very awkward."

P. Q. was in a very critical position. His son was one of his family and had therefore always been protected from his father's swearing. In this situation the father was just as emotionally disturbed as the man who takes his wife out of a public house, or starts a fight if bad language is heard.

Another collier of about the same age shared endless suggestive conversations with his workmates and when one of the younger men's sex-life comes under discussion – e.g. a youth might be 'kidded' because someone knows he has seduced a girl – he takes the same amoral attitude as the other men, an attitude which is more or less one of approval. But this same man spoke in violent tones of a young collier keeping his daughter out after 10.30 p.m. and described how he had caught them saying good night at the street corner and told the young man to be off at a reasonable time of night. His daughter had received a lecture on "not letting any-body take advantage of her" and the threat of a good hiding if she did not mend her ways.

For many reasons superstitions of all kinds have weakened over the past few generations, but nevertheless, there are remembered in Ashton certain beliefs concerning the pit and the sex-distinction which suggests that we are not exaggerating in suggesting a tendency to an almost ritual distinction. Ashton men and women were asked if they believed the story that in a small village in South Yorkshire miners had refused to go to work because a woman had crossed the pit-yard as they waited to descend. No surprise was expressed, and it was confirmed that the old miners in Ashton had held such views.

"Anybody will tell you that it was supposed to be bad luck for a collier to see a woman on his way to work in the morning, but, of course, nobody believes it now. The old miners used to go home if they saw a woman on their way and old A. B. he still stands by it, but he's a bit queer and makes it an excuse to have a shift off. Nobody believes it now. How can they, with women working in the canteens and con-ducting on the pit-buses?"

As the informant says, things are very different from the days
when miners got out of bed at 4 or 5 a.m. and walked a mile or
two in the dark to get to work. New realities have helped to break
down the superstitions. One other point is worthy of mention.
Provoked by the above discussion there followed a description of
other superstitions. It is still thought to be bad luck to turn back
once having set off to work, and one example was observed of a
man going to work without his snap and realizing this but refusing
to go back. His wife was sensible enough to realize what he had
done, and that he would not return, so she sent one of the children
after her husband with the food. Bad luck is also thought to result
if a garment is placed back to front or inside out and then corrected.
It must be left in the faulty position until the time comes to discard
it. Some of these notions are not peculiar to miners, but it is
interesting to note that in Ashton, and probably elsewhere, the
superstition about meeting women on the way to the pit is included
with these 'dangerous' and 'unlucky' notions.[1]

Our examples have been chosen as strongly as possible to suggest
that there exists a series of groups and activities in Ashton confined
to the mature males and that these groups and activities are separate
from and fundamentally *opposed* to the families of these men. As
many domestic disputes are caused by the young (and sometimes
the older) wife complaining of the time spent by her husband
with his mates as by disagreements and difficulties about money.
The money disputes themselves, of course, are essentially concerned with this question: in which sector of the community shall
the money be spent—in the family and the home, or 'in the Club'?
We have already quoted in the chapter on leisure the men who
recognized the women's attitude to 'the Club'—a very common
one—"if all the women's wishes for the club had been granted it
would have been blown into the middle of the Sahara desert long
before now".

---

[1] Superstition in Ashton is of little importance. One superstition which does persist and
is well included here as anywhere is a belief similar to the widespread custom of 'couvade'
in which the men of certain primitive societies go through the stages of childbirth while
their wives are actually going through labour and giving birth. When an Ashton woman
is pregnant her husband is expected to suffer from toothache, headache, stomach ache and
all manner of strange ills. If he complains he will be told that with his wife in her present
condition he ought to expect this.

Stages of development in the group of males

Indeed this antagonism between a man's wife and family on the one hand and his mates on the other even manifests itself before marriage. The youths in Ashton spend most of their leisure-time in groups of about half a dozen. Such a group will grow naturally out of schoolday friendships, perhaps with additions from workmates or those with whom sporting interests are shared. After his evening meal the youth of between 15 and 18 or 20 will walk down to the street corner or the cricket field or the youth club, wherever it is that his particular group is accustomed to gather. They will go to Calderford to the billiard saloon, go to a football match or play themselves, and at the week-end visit a dance together. Round about the age of 18 most of them will begin to drink beer, though very few of them are yet heavy drinkers. It is usual for them to spend an hour or two in the public house before a dance; on occasion, one of the members will become involved in a scuffle and his mates will come to his aid. Often this results in a full-scale fight. A group of Ashton youths will often become involved with another group at a dance-hall in a nearby town, and if they are defeated, or if there is no decisive conclusion, the groups may take steps to 'sort each other out' at a future date. Such fighting is not uncommon, and in the three towns of Ashton, Calderford and Castletown it is rare for a Saturday night to pass without at least one such affair.

Fighting together is only the extreme of solidarity which grows up in these groups of young men. For years on end they will continue to share their leisure-time and it is soon remarked on if one of the members begins to mix more with another group. But the strongest competitions for the attentions and time of the group's members is, of course, sex-interest. Between the ages of 15 and 18 the youths will have only casual girl-friends. "I don't bother with girls. Sometimes I'll pick one up at the dance on a Saturday, perhaps take her out on Sunday, and never see her again", was a typical comment. At this stage youths will be anxious for any sexual experience, and occasionally this will be accomplished, but they are not yet skilled in the approaches to girls, they are awkward and embarrassed in the presence of girls – a state of mind which shows itself in a blustering and jokingly aggressive attitude. At this stage they are trying hard to be men; any display of tenderness or affection seems to be regarded as 'soft'. If one of the group of young males at this early stage does begin to take a girl seriously he finds himself the butt of consistent 'kidding'. Only a convincing and jealousy-provoking claim that he is intimate with his girl-friend will save him from ridicule. For the most part concern with girls is thought 'soft' and 'cissified'.

Nearer the age of 20, of course, when the young men are beginning to develop serious sexual affairs, and their stories of success are more likely to be believed, the attraction of the opposite sex becomes a real danger to the solidarity of the group. Now there is even serious talk of marriage, and a lad is prepared to admit, albeit after severe pressure, that he is 'courting strong'. These love-affairs often effect the prolonged withdrawal of young men from their groups. If such a man meets his mates in the street they will chaff him, "Well, we'd ask thee to come for a pint, but I expect tha's off to get thy feet under t' table." It will be loudly remarked that his girl has obviously got him where she wants him – i.e. in no time he will be married. After the first few months of intensive wooing the love-affair either breaks off or becomes more pedestrian, with prospects of permanence. Eventually the young man begins to insist on one or two nights 'with the lads' and it is at this stage, *not necessarily after the marriage*, that the struggle for the interest and attention of the husband begins. A few of the quieter young men sever their connexions with the group entirely; indeed, this type of man will probably not have been so deeply involved as others in the first place.

The attraction of the company of his mates proves to be a thorn in the side of the typical miner's family from this time onwards. A man is told of the projected activities of the group and does not like to excuse himself on the grounds that he wants to be with his fiancée or wife. In such a case he will be greeted with derisive taunts. "By God, she's got tight hold o' thee!" He finds it easier to assert his oneness with the group, his manliness, by committing himself to them and later fighting out the question with his girl-friend or his wife. Several young men in Ashton, having spent their holidays with their fiancées in the first year of courtship, went out of their way to arrange for the girls to spend their holidays separately in the following year, taking the opportunity of a full week 'knocking about with the lads' which occasionally means picking up a strange girl. One young man of 23, even though his holidays coincided with those of his fiancée, booked her accommodation at Blackpool and paid for the whole of her holiday. "I'm not having her hanging round all week through my holidays!" When the young man married the same girl six months later he did not tell his friends of the wedding until only a few days before the actual ceremony.

Actions of the kind described above are the background to the widespread custom of the 'bachelor night' a custom kept up in Ashton and in various forms throughout our society. On the night before the wedding the man meets his male friends and they have a thoroughgoing 'night out', which normally amounts to visiting as many public houses as possible and drinking the maximum amount of beer. Some men will boast of sexual experiences on such occasions, and in some groups a man's friends do attempt to 'fix him up' in this respect. The wedding itself symbolizes the beginning of a period in which a man's participation in this group of friends will be institutionally restricted. This is seen as a 'loss of freedom', and the custom of 'bachelor night' is a protest.

Once married, a man will certainly find, and expect, restrictions on his extra-familial activities, but there are many influences at work which determine to what extent the breach with his friends will go. Again the consideration of extremes is most instructive. Several men were interviewed and observed over a period who had only come to live in Ashton on their marriage to Ashton girls. Very noticeably these men spent very little time and money drinking or sitting in the clubs. They are better savers, are more interested in their homes, and spend more time with their wives than do other men. There is little doubt that a very powerful influence on their behaviour has been the complete break on marriage with the friends of their youth, left in another town or village. These men do not appear to have developed full relationships with their wives by any means, but their lives in this respect show more possibilities than those of most other married men. Other men who come into this category of more responsible husbands, generally less spend-thrift, are those who because of individual temperament or perhaps some absorbing interest did not become very thoroughgoing members of the groups of young men.

Certainly one very important differentiating factor in the early twenties is the wage-packet. Up to the time of marriage the wage can and does have two opposite pulls. It creates the conditions for the group to really 'go to town' in the pursuance of its objectives, and as the members approach the early twenties, receiving bigger wages, they usually take to drinking more seriously. On the other hand a good wage and the possibility of saving gives a man more confidence and prospects of success in courting. After marriage only those members of the group who go into face-work can afford to both support a family and continue to 'knock about with the lads'. Some of the lower-paid men try to do both but the difficulties are great. The method of dividing the wage which we have already described is the means of ensuring the face-workers' ability to maintain both home and pleasure.

Although there are significant differences within the group of males, according to wage-levels, nevertheless a general tendency is clear for members of the group in the early years of marriage. For in these early married years in this period the group tends to disintegrate, though it rarely does so completely. No matter how the wage-packet is divided, the various ways in which the husband contributes to the household finances are certain to weaken his ability to behave as he did before marriage. In the years of building up a home and raising a family most men's participation in the activities of the group grows less; the men return to them in middle-age when their family has grown up and the difficulties are less. These groups to which they return are sometimes the same as the groups of their youth, sometimes they are more closely related to work groups. Whichever is the case, it is at this stage that the groups flourish most.

In Ashton the family as a unit is weakened by the existence of a series of institutions and practices which are the domain of adult miners in the town. For outstanding individuals the positions of leadership in various organizations, particularly the trade union branch and the town council, are the foci of extra-familial time and attention. The majority of men, however, take their part in the 'secret society' of males through the medium of the small groups of friends and workmates. It is in these groups that they devote themselves to those activities which are viewed as specifically male in character, and here too they express their ideas in that unrestrained way thought fit only for male company. Each of these small groups strengthens in its members those attitudes thought characteristically male. A man who gives way to his fiancée or his wife is a weakling. A man who is 'tight' with his money when the group are together will not gain prominence or favour with the other members of the group. Miners like a man who says what he thinks in no uncertain manner, shows fear of no one, and is a liberal spender.

Within the group, it is permissible to rely temporarily on the generosity of the others; a man who is short of money will go in the club and his mates will buy his drinks 'while Friday'. But this is a point of honour that each man 'pays his turn'. Even in the best-spirited of groups, disputes can arise at the end of an evening if someone is suspected of dodging his turn. No one says 'It's your turn now' except in jest; any man who has not yet paid for a round of drinks calls the waiter and 'gets 'em in'. The reciprocity in buying of drinks is really the hallmark of the groups. Every man in the groups must 'be a man'. One finds men with weak characters asserting themselves in such groups. One man, A. P., showed signs of extreme lack of self-control and ill-temper towards his wife and family, and appeared to be in every way the spoiled product of an over-fond mother, was noticeably over-demonstrative in his buying of drinks. When he had his pocket full, he insisted on buying more drinks than he could afford, and was very anxious to be thought a grand fellow. Behaviour in the groups of males, except for a few leaders of organizations, requires no great strength or quality of personality, only an ability to keep up with the rest. The activities of the group are non-creative and show no development through a man's lifetime. They call for no initiative,

15

only the capacity for enjoying oneself freely and companionably in the limited round of sport, conversation and drinking. Into such groups can be placed all married miners in Ashton with the exception of a few who have married in from outside and a small number of exceptional individuals. Also included are some non-miners, working-men who have grown up with the men who are now miners and have always shared their pursuits outside of working-hours. From their youth men are conditioned to enter this closed society of working-men. From boyhood into manhood the small groups in which males share their actions and thoughts maintain and strengthen the ideas of manliness being opposed to anything to do with girls and women, except in terms of sexual conquest. We have seen that this hold on the men lasts even through the years of most intensive relations with the opposite sex and is invariably reasserted completely after the early years of marriage.

For the intimate lives of Ashton miners and their wives the effects of such a powerful attraction outside the home for one partner can only have the effect of a strong disintegrating tendency. That the effects are not wholly disintegrating is only due to the acceptance of the existing state of affairs by women, who, although they protest intermittently (and in some cases continually) about the lack of attention and responsibility shown by their husbands, do not challenge the whole basis of their separate lives. Women may criticize the conduct of their husband, and they occasionally have strong feelings about the conduct of affairs thought to be exclusively the concern of men, but they never carry out decisive action. During a short three-day strike in Ashton women spoke often of the difficulties with which they would be faced in trying to run their families without their wages. The following conversation occurred in a public house – when the local trade union branch secretary entered.

"Why aren't you at work?"

"Our face is on strike," said the secretary.

"Do you want a good speaker?"

At this the woman serving behind the bar interjected, "You want to let the women go and speak, and then see if they stay on strike."

Another young wife, during the Winders' Strike of December 1952, said, "I saw two men with silly grins on their faces shouting to each other across the road about having a nice few days off work. I felt like telling them they'll be smiling on the other side of their face when they see their wage next week."

In general men do not strike irresponsibly, and at any strike meeting men will speak about the effects on their families, but there is in the remarks of both these women an expression of the feeling that the actions of their menfolk are to some extent determined by quite other considerations than those of their families' welfare, indeed that this welfare might be left to suffer at the whims of the men. It should be stressed that this is only a feeling. In any particular strike if it comes to the question of solidarity the wives usually defend their menfolk strongly; they share the sense of injustice felt by miners and support their husbands against the enemy.

THE PERSONAL RELATIONS OF HUSBAND AND WIFE

Against the background of traditional division of duties and responsibilities between husband and wife, and the separate sectors of social relations in which they participate, what kind of relationships normally exist among married couples in Ashton? As already intimated the great majority of women are successfully conditioned to the acceptance of their situation. Together with their real and very close ties with the children and the difficulty of their setting up independent units within the given economy, the ideas persisting among Ashton wives make up a whole system out of which any individual will find it difficult to break. 'Public opinion' condemns the woman who exposes her children by forsaking her husband, the wife herself knows the difficulties involved in such a step, and last but not least, she has been brought up in a society where she expects a state of affairs very little different from the one into which she enters at marriage. Our concern has not been the intensive psychological study of individual pairs of husbands and wives, but it is suggested that such individual problems will make sense only against the background of social structure and ideology which we have tried to outline.

The general features of the relationship of married couples in

Ashton are determined by certain social relations existing outside of the family and a whole system or structure within which the husband and wife are included. In such a situation the development of deep and intense personal relationships of an all-round character is highly improbable, and observations confirm the absence of any marriage corresponding to the ideals of romantic love and companionship. Many married couples seem to have no intimate understanding of each other; the only occasions on which they really approach each other is in bed, and sexual relations are apparently rarely satisfactory to both partners. Because of the divisions in activity and ideas between men and women, husband and wife tend to have little to talk about or to do together. It is, therefore, a common feature for no development or deepening of the husband-wife relationship to take place after the initial intensive sex-life of early marriage. Indeed those couples which seemed happiest were in the first year or two of marriage, when most problems were solved by going to bed. Apart from these (and perhaps a few highly exceptional cases, though these were not discovered) marriages in Ashton are a matter of 'carrying on' pure and simple. So long as the man works and gives his wife and family sufficient, and the woman uses the family's 'wage' wisely and gives her husband the few things he demands, the marriage will carry on.

Here in the Ashton family is a system of relationships torn by a major contradiction at its heart; husband and wife live separate, and in a sense, secret lives. Not only this, but the nature of the allotted spheres places women in a position which although they accept it, is more demanding and smacks of inferiority. And yet marriages in Ashton do hang together; disintegrations are rare. The tension inherent in each individual relationship may be eased or intensified by individual idiosyncrasies, and is, as we have seen, altered in emphasis by individual economic resources, and type of work, but the tension always exists as a social fact by virtue of the social structure of which all husbands and wives are part. The contradiction and conflict in the situation is not entirely overlaid and suppressed by the weight of ideology. The emotional effects on the participants will not allow of this.

Since the economy and prevailing social conditions do not allow for complete separation – the revolt of the wife – there can only be occasional outbursts – temporary rebellion by the wife (or in a disguised manner by the husband, for although he appears to be in a superior position, nevertheless the system militates against the fulfilment of all his emotional needs). Regularly or irregularly the wife will protest against her husband's failure to give her sufficient money or his lack of consideration for her. The quarrel resulting from one of these occasions, like any other discussion or argument in which tempers are raised between husband and wife, is called just a 'row'. It is not an exaggeration to say that 'the row' is an institution for the present-day family, at least in Ashton. Conditions 'external' to each individual family are responsible for tension in the family; those same conditions make it impossible for an alternative to be achieved by revolt against the whole structure of relations in the family; thus the conflict between husband and wife is turned in upon the family itself. The row is the conventional way of expressing the conflict. At the same time it is a release.

It is in rows and in the lack of display of affection that the nature of marital relations shows itself most affected by the social structure and its concentration of advantages for males. The whole life of the miner under the influence of his group of friends inhibits any display of tenderness and love in sexual relations. Men are reluctant to discuss with a stranger their personal marital relations: as much as they will say under pressure is that after a few years a man is bound to get fed up with his wife.

In the following example H. B. is more frank than many men, and certainly more promiscuous, but his and his wife's attitudes are nevertheless typical.

"H. B.'s wife (29), an attractive woman with two children, has this to say about her marriage: 'We've been married seven years. I get a regular wage from him. I've got a home, a fairly decent home, and two kids, I can keep clothed and fed. And it's that security that I want most of all. I know H. gets about and spends a lot that I don't get to know about, but I have enough to keep going – he used to be very romantic, but he's changed a lot. I'll go over and sit next to him when he comes in from work, and he'll say to me "Can't you find somewhere else to sit, and let a man read the paper in peace?" He just isn't lovable and I can't seem to do anything about it.'

"Her husband frankly says that he no longer gets any enjoyment out of sexual relations with his wife, and prefers extra-marital affairs. He says he thinks his wife should be satisfied with life:

" 'I bought her a new wireless, and then a television, and I hired a car for the holidays, and still she natters about what time I come in and where I've been, and can she have an extra shilling or two. There's no pleasing 'em.' "

The latter sentiment was echoed by an old collier, S. P., after telling two young men how he had to deal with his wife. "Take my tip, once you let one o' them get top side o' you, you've had it!"

The extremes of families which 'carry on' and nothing more, as if acknowledging the clash of interests between husband and family, are those of older workers who spend no time with their wives. A woman of 55 tells the following story:

"You'd think from when you come in here, with my husband (F. G. – a day-wage worker), that things were as nice as pie between us, but the fact is we never hardly speak to each other, and it's been like that for nearly 30 years. We quarrelled over him going out with another woman before I had Jean (eldest daughter, 29) and since then there's been nothing between us at all. This is home, he gives me money at the week-end, and I look after him and the home and the girls. He sleeps and eats here; if there are times when he sits in the house it's because he's broke; he never opens his mouth and neither do I. About five years ago, he brought home a woman about 15 years younger than him, introducing her to me, and said she was to stop. I gave her her tea–I thought if she's the cheek to shake hands I could give her her bloody tea!–but then my four daughters came home from work (then, 16, 19, 21, 24). As soon as they saw what was going on they threw her out of the door and when F. interfered they gave him the thrashing of his life, kicked him, thumped him, pulled his hair, and since then he's never tried it on. I've got too used to this carry-on; I know I ought to have left him years ago, but there's nothing I can do now. Where could I go?"

For the men, the pattern of reaction to the almost business-like arrangement of marital relations is a suppression of all apparent interest in 'affairs of the heart'. In their reading and film-going, and most important, in their conversation, they show a taste for sport, adventure, toughness and absence of 'sloppiness' and 'soft-ness'. Sex is something different from the relations between human beings, a matter of conquest and achievement, for the male indi-vidual. In such an ideology women can only be objects of lust, mothers and domestic servants. The sex-life of married couples shows the effect of this.

Very few women stated real satisfaction with their sex lives. In other cases women complained of their husband's selfishness in not considering the woman's complete satisfaction. The widespread practice of withdrawal as a measure of birth control can only detract from the likelihood of female orgasm. These conditions combine with the traditional reticence in open discussion and expression between the sexes in such matters to make many women feel 'cold' in their marital relations. In addition girls and young women have a general attitude towards sex which is very different from that of men. They express a preference for the 'romantic' in films and in cheap literature. A tear-jerking love-story will be referred to as a 'woman's film', a rip-roaring Western as a 'man's film'. Women read sixpenny 'romances' by the score. Men do not read them at all. Women's magazines help along this attitude on the part of the girls and women; they carry articles on cooking, dress and the home besides 'love stories'. These 'love stories' are often illustrated with suggestive and pas-sionate scenes but the text itself carries no description of love-making or direct reference to sexual desire. The 'affairs' of the heroes and heroines are full of the purest love and admiration, and their feelings towards one another are expressed in these terms rather than in terms of desire. It would seem that the preoccupa-tion of women with these stories is an expression of the lack of completeness in their lives. One woman described her 'ideal man' as a pipe-smoking, trilby-hatted, dark-moustached man. She did not say this in any spirit of criticism of her husband, a collier of average size with blond hair, a non-smoker and clean-shaven into the bargain.

The attitude of women towards their boy-friends and their prospective husbands is thus very different from the approach of the men, and reflects their position in the society. With exceptions, they do not have much conception of sexual satisfaction as an aim in itself. Rather they regard cohabitation as one part of the marriage contract which sometimes they enjoy and sometimes they do not. In the 'courting' stage, while the youth aims consciously at a sexual conquest, most girls have other criteria of desirability in a boy-friend and a future husband. Besides clinging ever closer to the 'romantic' notions of films, magazines and novelettes, there is another reaction to the disappointment of a woman in the 'romance' of her marriage. She takes over a garbled version of her husband's attitude towards sex as something 'in itself' and groups of women will often spend a great deal of time in chatter about the sexual habits of their husbands, or repeating dirty jokes and making suggestive remarks about each other. This is the other side of the coin to their preoccupation with the purity of the love-affairs appearing on the printed page. Incidentally, when men talk about sex they do not talk about their wives or the wives of anyone present, either in general or about some other person of whom they all know. Again this indicates a division between the family on the one hand and the desirable pursuits of males on the other.

In the past few years, i.e. in the period during which have grown up school-leavers since the war, certain developments suggest a change in the new generation of women. Open sexual interest was for the older generations confined to the groups of men and a few girls who were 'a bit free' – there always seem to have been a few of these in any village or part of a town.

Since the war the emphasis on 'sex' rather than 'love' and 'romance' has increased and become more open. Weekly magazines of a certain type are widely read by young women as well as men, and in these 'sex-appeal' is very deliberately cultivated. The trend in films and in the increasingly popular American pulp novelette is towards pornography and sex as part of a whole picture of violence. Women are as directly influenced by these developments as their brothers, boy-friends and husbands. A woman who was thirty in 1953 was very different in her attitudes, derived from her reading and film-going experiences, towards sex, and towards men, from her counterpart in adolescence in that year. All this can only tend to make the attitude of mind of girls towards sex approximate to that of the young men in the sense of seeing it more as something in and for itself. It is still, of course, presented as an affair in which the man is the conqueror.

In addition to this rather vague change in the generation now entering marriage there is the impact of new conditions, which threaten the persistence of the family structure and ideology to which the Ashton of old gave rise. Work for women is increasingly available both in Ashton and by means of an ever-improving transport system.[1] As a result of this, since the 1930's the great preponderance of males over females in the marriageable section of the community has disappeared. The prospect of independence is a small additional factor here. The most potent influence of all is the new-found comparative prosperity of the coalfields. Large numbers of young men show that stability and responsibility which is rarer among their fathers. It is doubtful if the higher wages can really change the pattern of family life in Ashton until there is an all-round development of social and cultural life. Once the family is well housed, it is difficult to see in the mining village of today how high wages are going to be spent. Nobody wants more than one television set. But at least the higher wages of the last fifteen years have given the new generation of mining families a fresh start.

CHILDREN IN THE ASHTON FAMILY

In the writing of recent decades the emphasis in the upbringing of children has been fairly and squarely placed on the development of personality, as determined essentially by the environment of the child's earliest years. For the ordinary parent, however, the impact of this theoretical tendency has not been great, and the two aspects which still receive most attention are the physical well-being of children and their 'education'. Character-formation and personality-development take place by a process accidental to the intentions of the parents, who concern themselves first of all with feeding, clothing and generally looking after the health of their children, and secondly with their formal education.

[1] See Chapter I, 'Place and People'.

Parents and the destiny of the child

Formal education is by now the prerogative of the schools, and the role of parents is confined to that of encouragement and the creation of the right conditions at home for their children to take advantage of their schooling. The physical well-being of children is still in the hands of parents, who are helped in this respect by family allowances of 8s. per week for every child after the first, by the provision of a public health system and by the provision of school meals. As for personality-development, a process which ordinary parents do not consider for its general aspect–they may say a child is bad-tempered or jealous, etc. etc.–it is outside our scope to consider this subject in detail. We simply note in passing that boys in Ashton are destined to be Ashton miners and girls to be the wives of these miners. Their personalities must first conform to the requirements of these roles in society. When they are adults they must be fitted to function in the ways we have described for Ashton men and women.

When we said that personalities developed accidentally to the wishes of parents we had in mind the real, everyday conditions of the child's life. Most important of all is the fact that to a great extent the attitude to children is the same as that towards the conduct of life generally in Ashton. It is essentially a day-to-day or 'hand-to-mouth' outlook. Intensive observation of parental discipline in a dozen families confirms this view. A second fact of major importance in this respect follows from the division of duties between husband and wife. In fact, *the bringing up of children is the job of the mothers.* This, together with the fact that bringing up children is in fact *looking after* the children each day, is at the root of the mechanism of personality-development in the early years.

In different families, other factors will assume more or less importance in modifying these basic conditions. Where kinsfolk live very near, the child is fortunate enough to be close to a number of adults besides the mother. Large families will give rise to a relationship between mother and child different from that produced by small families. If a woman has bad relations with her husband, the result will either be a development of cynicism and neglect of the children or a closer 'compensatory' attention to them. Exceptional parents may help their children on every way to take advantage of the opportunities available in education, and act as though they see the real possibility of a prospect of other work than the pit for their sons. All these additional factors work within the framework outlined; an undirected process, concentrated mainly on the immediate comfort and well-being of the child, and in the hands of the mother, rears boys to be miners and girls to be their wives.

Ninety-three Ashton persons selected at random were asked about their attitude towards encouraging a son to work in the mining industry. The vast majority, 66, said that they would not or did not encourage their sons to this step, many of them expressing themselves most emphatically. Many added, however, that they would allow the boy to choose for himself. This latter statement, however, is obviously significant, since despite the fact that parents apparently do not encourage their sons to be miners, the great majority of them do in fact become miners. Indeed, this question was only asked in order to confirm the statement heard many hundreds of times in the Yorkshire coalfields since 1945 – "I hope no son of mine ever works in the pit." When the same persons were asked if they thought it advantageous to have a grammar school education for both boys and girls 79 answered in the affirmative but a few added their opinion that a good education for girls was likely to be wasted since they 'only get married'. It is unlikely that this preference for grammar school education has any effect in reality. It is reliably learned that the mining areas are the best areas in all Yorkshire for selling such publications as *The Children's Encyclopaedia* – publications which cost from £12 to £15 and which parents buy because they want 'to do something' for the child's education. Yet only very few Ashton children actually get a grammar school education and very few develop interests broader than those of their parents whose interests can be satisfied by life in Ashton. One reason for this is that the parents themselves obviously can have no real idea of a child's requirements to advance in the educational system. In the competition for grammar school places the child of well-educated parents has inestimable advantages. His parents not only give direct instruction, help and advice, but, for example, they understand the need for good conditions for the child to do his or her work at home, and appreciate the relations between a tranquil emotional life and successful study. Working-class parents, with a growing number of exceptions, can only grope for such advantage.

However, it is as well to make this point. We have given two examples of the failure to coincide of parents' expressed wishes and the actual fate of their children. In each case an important consideration is the fact that the position of the parent in the social structure unfits him or her for the task of (a) holding his ideas very strongly or practically and (b) carrying them out. These two disadvantages, not by any means directly perceived or understood by all parents, reinforce that characteristic we have already described as 'basic'–the approach to life on only a day-to-day footing. Miners, like many other industrial workers, are often very cynical about 'book-learning' in anything to do directly with their own work and life. They recognize the value of book-learning for 'getting on' in school and as a training for professions out of their experience, but for their own lives, and in contact with their own people, they are sceptical about 'theory', so that any child not showing exceptional talent with 'books' at school is given little encouragement to study. The vocational training available in mining institutes after leaving school is seen as much more valuable.

It is in matters of discipline in the home that the 'day-to-day' attitude of parents is most clearly in evidence. Children are washed, dressed, given breakfast in the morning, and then placed in their routine activity for the day. Children under age are 'put out to play' or "given something to occupy themselves with"–in Ashton rubber comforters or 'dummies' are still widely used, often up to the age of 4 or 5. The older children are seen off to school. It is, of course, the mothers who carry out these tasks. Meals for the children are often taken with the father, but shift-work may interfere with the regularity of joint meals. At a certain time the children are washed, undressed and put to bed. Discipline in the house is usually directed against the child's failure to allow the daily routine to run smoothly. A close second category of offence is interference with adult activities. "Be quiet when somebody's trying to talk" is the acme of this attitude to children. It is very common indeed. It is true, in summary, to say that discipline is administered not with an eye to the development of the child but to the immediate needs of the parents and the household routine. A corollary of this is that punishment varies not so much with the seriousness of the offence as with the state of mind of the parent. A child will repeat the same slight offence a dozen times before receiving a tap or blow at the point where the father's or mother's patience is exhausted or the point at which the child's desires conflict with an express purpose of the parent.

This preoccupation on the part of the mother with the routine of her daily duties can only give rise to an attitude to the children which detracts from any serious and truly helpful handling of their development. A mother may be proud of the growth of her children, she may earnestly enquire after their progress in school, but her own relations with her children consist of *dealing* with them, *coping* with them. Mothers in Ashton, as anywhere else, show great love for their children; proudly as they grow they tell their neighbour of their concern for feeding them and keeping them happy; they appear to understand the need of the child for maternal affection. In the prevailing conditions this love is principally expressed in spending as much as possible on clothes and toys for the children, often uneconomically. It does not manifest itself in a serious and detailed consideration of the development and problems of each child and in a plan of action to bring up the child. The simple fact is that mothers, who have been given the sole responsibility for the early years of childhood, are not educated in the necessary knowledge for this task, whether or not one gives any weight to the 'natural' power of maternal feelings. *Looking after* the children is the phrase most commonly used for a woman's duties towards her children, and this describes well the mechanical view taken of their tasks by most mothers.

The much closer relation of the children with the mother than with the father also has important consequences which are closely interrelated with those already discussed. Important among women is a sort of rivalry over their children. Women say of each other that they are clean or not so clean in their household duties, and in one street gas cooking-stoves or kitchen cabinets may spread as much for prestige value as for anything else. Similarly with the children : at Christmas and Whitsuntide a woman buys clothes for her children not only because they need them but because she does not want to feel inferior to other mothers who do this. A woman who clothes her children shabbily is talked about. When women gossip together they describe the Christmas presents they are buying for their children.

Women are denied participation in those activities whereby men achieve success or reputation. They definitely try to assert their individual worth among other women by doing the job of motherhood as well as or better than their neighbours. In fact this means showing the outward signs – new clothes, new toys, well-fed children. It is by these standards that a mother is immediately judged. The child is in the dangerous position of being a status-object for the mother.

The mother's preoccupation with the household routine and her narrowness of outlook which her social position produces, together with the almost perpetual absence of the father from the domestic scene, means that the children receive very little detailed and consistent attention from their parents. One soon observes that in Ashton only the exceptional parents can really give any of their time to talking and playing with their children, seriously teaching them new things, introducing them to new worlds. This task is left principally to the other children with whom the child mixes, to the school, and to the films and comics. The neglect of this side of the child's development is part of that 'day-to-day' attitude which is everywhere so important in the life of Ashton men and women.

We have seen very briefly that some of the main features of Ashton's social structure are strongly reflected in the upbringing of children; among those features are the confinement of the mother to the home, the isolation of the father from the home, and the ideas and attitudes of both father and mother consequent on their social roles. These form the framework of the child's development. They serve to give a bias to the life of Ashton children in favour of the static and restricted way of life already existing there. In other words, that system of roles and responsibilities which is the family of a man and wife and their children is a strongly formative character in rearing children who will perpetuate Ashton as it is. The family is thus a conservative force, a force for inertia in the culture of this community.

It is proposed to deal very briefly with two aspects of the upbringing of children in order to illustrate this thesis. These are (a) the difference in parental attitudes and practices concerning their sons on the one hand and their daughters on the other, and (b) the mother-child relationship.

Socialization and the sex-divison

Although the father does not normally spend much time with his children there is usually on his part a strongly marked favouritism for his sons. At a very early age he will occasionally play with his son, yet with his daughters he spends no time at all; it is a common sight to see the small daughter in the family standing mournfully aside while the father and son play together, and then be chastised for 'moaning' when she tries to attract attention. If the father can be prevailed upon to take one of the children 'down the garden' or to the cinema in the afternoon when he is on nightshift, he is more likely to take his small son. This favouritism applies later in another sense. Parents are much more interested in the educational progress of their sons than of their daughters, many of them regarding education for women as a waste of time, since they are destined to spend their time as mothers and housewives. For the boy who does not gain entrance to the grammar school at the age of 11, this interest ceases, but there will usually be some concern about what sort of a job he is going to take, some concern for his future. For the girl this is not true at all.

When toys are bought for the children there is not surprisingly a bias towards sex-typing. For the little girl there will be dolls, prams, toys like washing-machines, and so on; for the boy model trains or aircraft, a gun, a 'cowboy' set. This choice is made for the child before he or she will assert a preference in any direction, but, of course, once set on one path his or her own preference will follow the lines thought 'natural' for a boy or a girl.[1] From an early age the little girl will think it great fun to share the household tasks, but such recreation soon breaks off with the boys, who, under the influence of their fathers and friends, begin to pursue the more 'male' activities, i.e. interest in all kinds of sport, spending time with groups of boys rather than in the home. During adolescence the girl will be expected to help her mother in the menial tasks of the household, but the son usually gets off scot free like his father. When a young man marries he has already in his own home become accustomed to seeing the women do the housework while he is at work or at play. He grows into the traditional sexual division of labour inside his own home. He models himself on his father, even though his father is not very often in the house. A young woman knows the life of her mother; she grows up into the idea of how a good woman should behave and should go about her work. Like her mother she will know that she will be praised for 'looking after' her husband well, making a pleasant and easy-going homestead as a contrast to his arduous toil at the pit. More young women take up industrial work than ever before in Ashton but even among these the future looked forward to is marriage and a family as the end of work. Women half-jokingly say that they look forward to being married so that they can stop working. They build up for themselves dreams of ideal little homes on the model of the women's magazines. In reality, they accept, in the very great majority, the ideology of the sex-division in Ashton.

The girl's acceptance of her future as a housewife, with no other prospects of advancement, no notion of social participation outside the home, fits in very well with the more subtle aspects of what is in our society considered 'natural' for the feminine temperament. Thus a woman is likely to have little initiative beyond the fulfilment of her household duties; she knows that she will work within the limits of her husband's wage-standard. Within these

[1] Comparative social anthropology has made it quite clear that the behaviour and temperament thought natural for males and females are only the product of particular social and cultural forms. Like any other culture, that of Ashton types its children according to the roles of the sexes in Ashton itself. No doubt it is not very different from British society at large.

limits she will have responsibility, she must exercise patience, restraint, passivity in the face of hardship. Like her mother, subordinated to the needs of the men in her natal household, she will develop the characteristics of submissiveness and stolidity.

For the boys the opposite will apply. Certainly life in Ashton does not give scope for any great creative initiative on the part of men, although a few exceptional boys will advance in the educational system and, very rarely, in business. But within the given framework of Ashton and its social structure, i.e. in their *real* relation to their womenfolk, rather than in terms of any wider comparisons, the men of Ashton are the real units of the social structure, it is they who participate in the culture to the widest degree; they who may venture strong expression of opinion on those subjects which are discussed. In their homes boys and young men develop the idea and feeling that this is natural, and they take on the requisite personality-characteristics by imitation of their fathers and simply by fitting into each situation as they are obviously expected to do. Their parents encourage them to be 'tough'; if a small boy comes in and complains of being struck by another, he will be as likely as not turned out and instructed to return the compliment. If he swears at his mother when very young he is more likely to raise a laugh than to provoke chastisement. At an early age he will be taken to the football match with his father. From the beginning he grows into the sex-division of the household, and looks forward to the day when, like his father, he will be working. As we have seen, his adolescence sees the formation of those groups of young men which are so important in Ashton's social life. Often they have elements of continuity with the play groups of schooldays. In these very early groups there grows up the 'ethos' of what it is to be 'one of the lads' as opposed to the girls. All 'softness', gentleness, and sentimentality are crushed out in these groups.

Miners and their mothers

However, among miners in Ashton one trace of sentimentality does assert itself despite the influence of the group of males, and this is the miner's attitude to his mother. At first sight perhaps anomalous in the context of social structure we have outlined, this feature is in fact very easily shown to be closely related to and dependent on some of the cardinal features of the social structure. Men normally remain attached strongly to their mothers all their lives, visiting them regularly and helping them when they can. If the subject were directly raised, a man would say that he thought his wife a grand woman, but of his mother he is always prepared to give unstinted and uninhibited praise, often without encouragement. In this single case he seems able to speak of love without embarrassment. To the outsider this all seems rather maudlin as he almost waits for tears to come to the speaker's eyes. In a way this insulation of the mother from the rough and ready aspects of the miner's personality reflects that division between the family and the rest of the social structure. In similar fashion sex may be discussed in a group of men, but never do they mention their wives in such a conversation.

That reference to the mother is more freely in terms of feelings of love and affection needs to be explained to a great extent by the psychologist, but we can discern the social factors at the root of the attachment to the mother. We have spoken of the favouritism shown to boys by the father, but this must be seen against the background of one essential fact, viz. that of the home revolving around the mother. It is she who has the responsibility of every day and every week caring for the children, ministering to their desires. It is she to whom they turn when they are hurt or upset; she is the one who provides comfort in all situations when the child is ill at ease. She is the provider of all things, and this is true for a long, long time after weaning. In fact until the children are working, she can only do well her job of 'looking after' them by self-abnegation and sacrifice of her time and effort. Now the present generation of adult males has seen this self-sacrifice *to an extraordinary degree*. These are the men who were brought up in the 'hungry thirties'. When they describe these 'bad old days'[1] they inevitably dilate on the courage and strength of the mothers of

---

[1] The favourite comment of Ashton miners (no doubt not original) on the 'good old days' runs something like this:

"Things aren't what they used to be."

"No and they never bloody well were!"

that time. Habits formed in these difficult days still persist today. Many middle-aged and older mothers serve a heavy meal for their family and are content with a mere snack themselves. Over the years they have become used to the theory and practice of the mother giving herself to her home and her family. Many a man speaks of his mother's devotion to the children's welfare when their fathers were bringing home next to nothing on a Friday. Many a mother admits eventually that but for the children she might have found it impossible to maintain the tie with her husband.

Now there is no doubt that most men in the 'thirties' were extremely concerned about the plight of their families. Yet this concern was not expressed as openly and fully as it might have been. One expression of defiance of the conditions of depression was the retention of at least a semblance of the pleasures of good working-men. Such an attitude was reinforced by the values of the groups in which the men moved. Again, in these groups little value was placed on a man's devotion to his home. The fathers, then, even when they were desperately concerned for the welfare of their children, often disguised this strong feeling, keeping up a pretence of manly unconcern with domestic affairs. Thus they might completely fulfil their financial obligations to the household, but they would not often give that whole-hearted co-operation of the spirit which their wives would have found so invaluable. In their groups of friends and workmates even in the days of unemployment they kept up their interests in sport and gambling and they used the clubs even though they could not possibly spend a great deal. The sons reared in these families grew to see this as manly behaviour. As they grew into adolescence and manhood they spurned all show of affection and attachment to the home, and all connexion with womanly and childish things. It was manly to 'knock about with the lads', to expect to have everything 'laid on' at home. When in the groups it was manly to appear carefree with money, happy-go-lucky – 'one of the lads'. In growing up therefore, young men not only grew away from their mothers – this, of course, is inevitable and universal – they grew up into an antagonistic position without ever having cause to reject the services given to them over years of sacrifice and hardship by their mothers. Only in later life have they come to the full realization of their debt. Perhaps having children themselves has taught them something of the self-sacrifice required of a parent, as against the dependence of the child, but this is doubtful, since the husband has little of the responsibility. At any rate the young miner of today in Ashton seems to realize, now that he is an adult, just how much his mother had to give to raise him to manhood. He seems to feel some sort of guilt for his and his father's rejection of his mother.

It is within the framework of these social experiences that an explanation will be found for the generality of openly expressed attachment to the mother in Ashton. This love for the mother is not paralleled in attitudes towards the father. Many young miners, while praising their mothers, are critical of the irresponsibility of their own fathers and others of their generation. Among these younger men is developing a more responsible and far-seeing type of husband. When these young men hear some of the old colliers speaking roughly of their wives and asserting the justification for their spending so much time away from home, they will say that such attitudes are typical of the older generation rather than the younger. Some of the older men say exactly the same about the younger men. Both are right; both are wrong. In both groups there are men responsible and irresponsible towards their families. It does seem clear that although higher wages have been used, naturally enough, largely to perpetuate the pre-existing habits and pleasures of the miners, nevertheless there is more and more scope for marriages to have a fair start and develop along the lines of a partnership. For such a tendency to become dominant, it is clear that the social structure of Ashton must be radically changed. Within the present framework of a severe lack of employment facilities for women, the domination of local industry by coal-mining, the persistence of industrial relations which are coals on the fire of antagonistic attitudes in work and of cynicism in general, and the consequent wide gap in social status between men and women, the family will not be much affected by higher wages, family allowances or a few more places in grammar schools. All these changes are absorbed in a strongly established culture, a culture firmly embedded in social relations built up over nearly a century around the local colliery and within a period of history which has had direct and unmistakable impact on the miners of Great Britain. Family life as lived in Ashton is only the product of this culture; its members are only the participants in that social structure. Clearly the function of the family is as a mechanism for perpetuating the social structure, not only in terms of biological reproduction, but in terms of the production of the social personalities required by such a community as Ashton.

Conclusions

No attempt has been made in this book to describe completely the social life of Ashton men and women. Many important spheres of their action and thought have been entirely omitted, and those aspects which have been touched upon could have been approached from points of view other than our own. Such points of view would have brought out different sorts of facts from those which have emerged in our account. Our aim has been a limited one. We have taken three important formative influences on Ashton's social life – work, leisure, and the family – and tried to show that the type of approach used in modern social anthropology can expose significant features and interrelations in the material.

Certainly the work of social anthropology has so far with few exceptions been applied to primitive and to peasant societies, so that it might be feared to be inapplicable to the social structure of modern Britain. An anthropologist returning from a comparatively isolated and small-scale society was expected to provide a full and clear account of at least the main lines of economy, social structure, political organization, ritual and ideology. The field-monograph, typical product of modern functional anthropology, demonstrated at its best the consistent functional interrelations of all these aspects of a society's culture. But the anthropologist who studies his own society cannot do this. From the study of one community he cannot possibly produce a clear picture of the workings of the whole society and its culture. Normally the community he chooses contains representatives of only certain sections of the total society. It might by chance be a microcosm of the whole society; it might be a microcosm in one sense, and not in others; it might be entirely misleading to draw conclusions about the whole society from the study of one English community. To what extent the facts discovered and the relationship established are generally valid for the society depends not only on a demonstration of their internal consistency, but on a convincing placing of them in the context of national affairs and developments.

The economist in studying work in Ashton would have collected figures on production and income, on productivity and expenditure, presumably over a period of time. Some sociologists a study of trade unionism would have devoted more time and space than we have done to the formal structure of the union and negotiating bodies, their regularity and recruitment, their relation to general theories of large-scale organization and bureaucracy. A social pyschologist would have pursued different problems in the analysis of family life, showing the precise manner in which individual personality is moulded into Ashton's culture by the family life of Ashton. A demographer would have collected important data on the size of families, their family-building habits, the rates of birth and mortality.

Facts of all these kinds and many more are necessary for a thoroughgoing sociological understanding of our society or any part of it; the task demands large-scale planning, expenditure and co-operations between many disciplines. Meanwhile limited studies are undertaken, and they must recognize their limitations.

If the anthropologists' contribution to the understanding of our society can only be a partial one, it is nevertheless an important and necessary one. The stress laid by anthropological field-studies on the interrelations of a society's activities shows through also in the study of a modern community. Although Ashton's life in its many aspects is influenced by innumerable factors, great and small, which derive from 'outside' of itself, the principal lines of Ashton's institutions show an inner consistency and structure one with another. The foregoing chapters suggest that this consistency derives from the social relations of work in the coalmining industry in Ashton. For a study of Ashton these social relations must be taken as given. Their specific character could only be fully elaborated in a study of the national economy, putting forward all the facts about the relation of coal to that economy, the relative position of the miner in the economic and occupational structure, now and in the past. We have touched on these large problems only in so far as they affect our material, but clearly they are of vital importance.

The consequences of Ashton's population being a mineworking population are in evidence at different levels of action and thought in the community. In the sphere of the trade union, the strongest and most persistent of Ashton's associations, this influence is clear and unequivocal, so that cleavages among the men deriving from their types of work and income run through the organization of the union. Evidence is put forward to the effect that the union's own development as an organization, which has emphasized these divisions, has led to a situation of conflict between the needs and demands of the men and their representatives in the union. On the one hand, at the level of a community like Ashton, there are the 'informal' social relations of the pit. On the other the 'formal' bureaucratic structure of the union, with its full-time employees and close relations with the employing body. These conflict, not only in the minds of Ashton miners, but in practice, and this clash is most apparent at the structural meeting-point of the two sets of social relations, the branch committee. The men on this committee are still rooted in their work and in the community, but they are also on the first step of the ladder of the union structure, already entering into the kind of relations enjoined on paid officials, under considerable pressure from the hierarchy above them. This peculiar position explains their often contradictory behaviour.

Leisure institutions are influenced in a different way by the character of the mining industry. Because of a tradition in Ashton of a lack of facilities for female employment, because of the exclusively male character of mining, and because of the fact that in Ashton coalmining provides employment for the vast majority of working-men, there is an important division between men and women in social life. Institutional leisure activities are predominantly for males, and there is virtual or definite exclusion of women from many social activities. If the form of leisure activity is a consequence of mining, so is its content. We have suggested that the essentially frivolous character of leisure in Ashton is closely related to the insecurity, both physical and social, produced in the past and present by coalmining as an occupation in Britain.

In the discussion of family relations, it might be supposed that these intimate relations would be farthest removed from the influence of the pit, and certainly it would be foolhardy to attempt to trace such influences in every detail of behaviour. Yet the characteristics which Ashton families have in common, and the general lines of difference between types of family, are quite clearly very well adapted to the social and cultural requirements of Ashton as a coalmining community. These general characteristics, beyond, of course, the fact that each family unit is one unified by ties of cohabitation and descent, are not a simple matter of the family 'conforming' to the culture in a simple way. Certainly the wage-packet and the regularity of employment are a basic framework for family life, and set limits upon it whose rigidity must not be underestimated. But just as significant is the fact that the social life of Ashton creates groupings which cut across the individual family rather than demanding from it a simple conformity. In particular the sharp cultural division between the sexes, and the attitudes consequent upon it, run right through the community and produce tension within the family itself. The demands made on behaviour by this division are in conflict with the demands of the life of the family as a unit. At certain levels, this conflict is strong enough in its reaction to determine the division of the family wage in a decisive way. Our chapter on the family is a brief account of the working-out of these extra-familial influences in family life itself. The fact clearly emerges that the family can be understood in relation to other social units in which its members participate.

Finally, two points must be emphasized. A community study of this kind might give two impressions which are certainly not intended by the writers. The demonstration of consistency between different aspects of Ashton's life is not meant to suggest independence of external factors. The principal reason for our neglect of the major question of social change, for example, was the clear fact that in a community which is part of a wider economy and culture, the sources of change are operating at a higher level than the functioning of the community itself. In addition, while we are convinced that in many respects Ashton is typical of mining communities and of the industrial working class generally, research is necessary to establish the varieties of industrial community-life in Britain, and the sources of these variations. Clearly then, we hope that this book will be a small part of a field of research which is very necessary but which has hitherto been neglected.

Index